OXFORD MASTER SERIES IN CONDENSED MATTER PHYSICS

OXFORD MASTER SERIES IN CONDENSED MATTER PHYSICS

The Oxford Master Series in Condensed Matter Physics is designed for final year undergraduate and beginning graduate students in physics and related disciplines. It has been driven by a perceived gap in the literature today. While basic undergraduate condensed matter physics texts often show little or no connection with the huge explosion of research in condensed matter physics over the last two decades, more advanced and specialized texts tend to be rather daunting for students. In this series, all topics and their consequences are treated at a simple level, while pointers to recent developments are provided at various stages. The emphasis in on clear physical principles of symmetry, quantum mechanics, and electromagnetism which underlie the whole field. At the same time, the subjects are related to real measurements and to the experimental techniques and devices currently used by physicists in academe and industry.

Books in this series are written as course books, and include ample tutorial material, examples, illustrations, revision points, and problem sets. They can likewise be used as preparation for students starting a doctorate in condensed matter physics and related fields (e.g. in the fields of semiconductor devices, opto-electronic devices, or magnetic materials), or for recent graduates starting research in one of these fields in industry.

M. T. Dove: *Structure and dynamics*
J. Singleton: *Band theory and electronic properties of solids*
A. M. Fox: *Optical properties of solids*
S. J. Blundell: *Magnetism in condensed matter*
J. F. Annett: *Superconductivity*
R. A. L. Jones: *Soft condensed matter*

Band Theory and Electronic Properties of Solids

JOHN SINGLETON

Department of Physics
University of Oxford

UNIVERSITY PRESS

OXFORD
UNIVERSITY PRESS

Great Clarendon Street, Oxford OX2 6DP

Oxford University Press is a department of the University of Oxford.
It furthers the University's objective of excellence in research,
scholarship, and education by publishing worldwide in

Oxford New York

Auckland Cape Town Dar es Salaam Hong Kong Karachi
Kuala Lumpur Madrid Melbourne Mexico City Nairobi
New Delhi Shanghai Taipei Toronto

With offices in

Argentina Austria Brazil Chile Czech Republic France Greece
Guatemala Hungary Italy Japan Poland Portugal Singapore
South Korea Switzerland Thailand Turkey Ukraine Vietnam

Oxford is a registered trade mark of Oxford University Press
in the UK and in certain other countries

Published in the United States
by Oxford University Press Inc., New York

© John Singleton, 2001

The moral rights of the author have been asserted
Database right Oxford University Press (maker)

First published 2001
Reprinted 2003, 2004, 2006, 2007, 2008, 2009, 2010, 2011, 2012 (twice), 2014

All rights reserved. No part of this publication may be reproduced,
stored in a retrieval system, or transmitted, in any form or by any means,
without the prior permission in writing of Oxford University Press,
or as expressly permitted by law, or under terms agreed with the appropriate
reprographics rights organization. Enquiries concerning reproduction
outside the scope of the above should be sent to the Rights Department,
Oxford University Press, at the address above

You must not circulate this book in any other binding or cover
and you must impose this same condition on any acquirer

British Library Cataloguing in Publication Data
Data available

Library of Congress Cataloging in Publication Data

Singleton, John, 1960 Dec. 11-
Band theory and electronic properties of solids / John Singleton.
p. cm. — (Oxford master series in condensed matter physics)
1. Energy-band theory of solids. 2. Solids—Electric properties. I. Title. II. Series.

QC176.8E4 S46 2001   530.4′12–dc21   2001032856

ISBN 978 0 19 850644 7 (paperback)
ISBN 978 0 19 850645 4 (hardback)

14

Printed in Great Britain
on acid-free paper by Clays Ltd, St Ives plc

Preface

This book covers the important topic of band theory and electronic properties of materials. It is intended to be used by final-year undergraduates and first-year graduate students studying condensed matter physics as part of a physics or engineering degree. It may also be used as preparatory material for students starting a doctorate in condensed matter physics or semiconductor devices, or for recent graduates starting research in these fields in industry.

Why does this book exist?

Yet another book on the electronic properties of solids requires some explanation. In teaching final-year undergraduates and first-year graduate students, I have become frustrated with the scarcity of *general* texts which cover a wide range of material synoptically. Students at this point in their careers often have to embark on research projects, extended essays, literature searches, and so on, in very diverse topics; they tend to dive straight into a *specialised* text book covering one particular topic (one-dimensional solids, impurities in semiconductors etc.) and to forget that this topic is part of a wider whole. The outcome can be a blinkered attitude in which connections are missed, wheels are reinvented and so on.

Secondly, the 'old warhorses' by Kittel and Ashcroft and Mermin, although fine in many respects, have little connection with the huge explosion of research in condensed matter physics over the last 20 years. The leap from many such texts to current research topics is enormous; students, seeing illustrative data from the 1950s and 1960s are often convinced that condensed matter physics is no longer an active, glamorous research field.

Thirdly, there is a perceived gap between 'undergraduate' condensed matter physics texts, which often flounder at length in one dimension, concealing the wood with trees, and more advanced books such as Ashcroft and Mermin, which is often rather daunting for undergraduates.

I therefore planned a book which would treat band theory and its consequences at a simple level, but in three dimensions from the start, and which would provide pointers to recent developments. The book would give an overview of the field, suggesting literature which provided various routes into current research topics.

The task was made easier when the book became part of the wider *Oxford Master Series in Condensed Matter Physics*, a collection of six linked textbooks covering virtually all areas of the subject. Topics, such as superconductivity, which could lead to very lengthy diversions, could therefore be left to someone else, leaving what I hope is a much more focused book.

Background

Band theory is evident all around us, and yet is one of the most stringent tests of quantum mechanics. This book is therefore an attempt to reveal in a qualitative fashion how band theory leads to the everyday properties of materials around us.[1] It also provides many of the ideas and vocabulary necessary to understand the electronic, optical and structural properties of the materials met in science and technology.

[1] For example, silicon, calcite and copper all contain similar densities of electrons, and yet they have very different properties, all inexplicable without quantum mechanics.

An understanding of band theory has also led to an extraordinary burgeoning of solid state technology; the average resident of Europe, Japan or the USA interacts daily with hundreds of devices containing semiconductor logic, memory or optoelectronics. A huge amount of research world-wide is devoted to optimising and designing such devices. This book therefore aims to provide the vocabulary and quantum-mechanical training necessary to understand electronic, optoelectronic and other solid-state devices. Whilst not treating the operation of devices in detail, the book introduces holes, effective masses, doping, the use of reduced dimensionality, excitons etc. at the level required for advanced texts on the fundamental principles of device design.

The use of the book in a structured course

The book was designed to accompany a course of the same name in Oxford, and it may be useful to give a brief description of this process. The course comprises 12–16 lectures and three or four problems classes. Chapters 1–8 of the book, corresponding to perhaps the first 70 % of the lectures and first two classes, are intended to familiarise the student with the ideas and terminology of electronic bandstructure (e.g. bands, holes, Fermi surface, effective mass tensor, Landau quantisation, quasiparticles etc.). Thus equipped, the student can proceed to the rest of the book, corresponding to the final lectures and third (and fourth) classes, which deal with electron transport; i.e. the concepts derived in the first part are used in the second to study e.g. the transport of heat and electricity in metals and semiconductors, simple electronic devices and the quantum Hall effect.

In all of this, it should be emphasised that the exercises are very much an integral part of the book; they continue derivations and proofs started in the text, illustrate the new concepts encountered, and allow the student to become familiar with the typical sizes of important parameters. To gain maximum benefit from the book, the reader should attempt a substantial proportion of the exercises. Students vary greatly in ability, and I have included an appendix containing hints on how to tackle the more taxing problems, plus some numerical answers.

I have tended to assume that the reader will have already encountered some of the ideas of condensed matter physics, at the level of (say) *The Solid State* by Harry Rosenberg (OUP, Third Edition). However, this is not strictly necessary, and I hope that the appendices make the work reasonably self-contained, or at least provide a useful revision of related material.

In order to leave space for recent developments, the Drude model and the introduction of quantum statistics are treated synoptically, with an emphasis on the underlying assumptions and failings in Chapter 1. The need for a deeper understanding of bandstructure is thereby illustrated. Chapters 2–5 then introduce Bloch's theorem fairly rigorously in three dimensions, at a level slightly below that found in Ashcroft and Mermin; by this stage of their career, I believe that students are ready for such a rigorous treatment with 'no cheating'. The book then considers (in three dimensions) two tractable limits of Bloch's theorem, a very weak periodic potential and a very strong periodic potential (so strong that the electrons can hardly move from atom to atom). It is demonstrated that both extreme limits give rise to *bands*, with *band gaps* between them. In both extreme cases, the bands are qualitatively very similar; i.e. real potentials, which must lie somewhere between the two extremes, must also give rise to qualitatively similar bands and band gaps. The ideas of effective masses, band shapes and bandwidths are related to the *real space* arrangement of the atoms or molecules making up the substance in a qualitative way using the tight-binding model.

Having introduced the ideas of effective masses and holes, semiconductor bands are introduced in more detail, along with the idea of artificial structures such as superlattices and quantum wells, layered organic substances and oxides; again, all are qualitatively understandable using the ideas of the tight-binding model. In view of the presence of the book on soft condensed matter in the Master series, I did not consider it necessary to treat amorphous solids.

Most of the available condensed matter physics texts provide a very poor coverage of developments in the last twenty years. From Chapter 7 onwards I mention some of the current 'hot topics', and show that they can be understood using the techniques developed thus far in the book. In other words, by this stage the students have acquired enough background information to grasp the ideas behind recent research. Moreover, in illustrating examples of phenomena such as the de Haas–van Alphen effect, I have used recent experiments. For example, a crystalline organic metal is used as a 'case study' in Chapter 8. This is no accident; since 1998, more papers have been written on the physics of these materials than on the high-T_c cuprates.[2] However, in five years, the hot topics of condensed matter physics could be very different; if subsequent editions of this book are ever produced, I should expect to use different materials as my examples.[3]

I have tried wherever possible to use the standard symbols for quantities encountered in research papers and other fields, so that transferring from this book to other works is as painless as possible. Inevitably, this has led to one or two conflicting definitions,[4] but these are *always* made clear by the context. A list of the more commonly used symbols is given in one of the appendices.

In under three hundred pages, a book cannot be comprehensive.[5] Each chapter therefore ends with a section pointing out more detailed or differing treatments for the ambitious and simpler alternatives for the challenged. Relatively recent review articles and sources of supplementary data are also given; in this way, I hope that the book can function as a route to deeper, more specialised study.[6]

[2] I am grateful to Paul Chaikin of Princeton for this information.

[3] When reading these chapters, one of my older colleagues said 'but what about the seminal experiments on the magnetisation of metals in the 1930s?' My answer was that, whilst these examples are historically interesting, and worthy of considerable respect, they are not representative of this thriving field of research today. There are innumerable pictures of the oscillations in Bi and Cu in existing books, to which the student is referred if they are so inclined.

[4] For example, n is both the electron density and the quantum number of the Bohr atom; p is both the hole density and the momentum.

[5] As another John wrote 'And there are also many other things ..., the which, if they should be written every one, I suppose that even the world itself could not contain the books that should be written'.

[6] I believe that it is a mistake to make a book *too* comprehensive (e.g. some of the books which purport to contain the whole of 'University Physics'). This discourages the reader from seeking enlightenment elsewhere, and can eventually lead to the belief that one doesn't need any other text book or paper to 'know physics'. Gaining experience in sifting information, doing literature searches and seeking connections is invaluable and helps to develop the student's critical skills.

Acknowledgements

This book has been greatly improved by the suggestions of numerous friends and colleagues. In Oxford, Arzhang Ardavan, Steve Blundell, Geoff Brooker, Anne-Katrin Klehe and Mike Leask have read the various drafts of the book, making many useful comments, donating figures, doing the problems and correcting errors. Mike Glazer provided a critique of the appendices, ensuring (I hope) that I do not infuriate crystallographers and others. Elsewhere, Greg Boebinger (Los Alamos National Laboratory), Martyn Chamberlain (Engineering, Leeds University), Helen Fretwell (Physics, University of Wales), Marco Grioni (EPF, Lausanne), Neil Harrison (Los Alamos National Laboratory), Steve Hill (Montana State University, Bozeman), David Leadley (Physics, University of Warwick) and Jos Perenboom (Physics, Katholieke Universiteit Nijmegen) made excellent suggestions which I hope will make the book more widely applicable.

I have used various incarnations of the book in teaching undergraduate and graduate students in Oxford and elsewhere. I should like to thank the many victims of this process who have pointed out errors, typing mistakes and incomprehensibilities. Too much space would be consumed if I thanked *everyone* by name, but particularly enthusiastic or devastating in their suggestions were Lily Childress, Carol Gardiner, Paul Goddard, Eleanore Hodby, Lesley Judge, Brendon Lovett, Eleanore Lyons and Amalia Coldea, to whom I am very grateful. In spite of this enormous effort by others, some errors will inevitably remain, and these are entirely my responsibility.[7]

A substantial number of the figures in this book were drawn by Irmgard Smith, who made much of my rather wobbly initial sketches; I am very grateful to her for this enormous effort, which allowed me to spend more time on the text and the other figures. I should also like to thank Karen Dalton, Phil Gee, Phil Klipstein, Robin Nicholas and John Ward, who supplied data or images for use in the book. Mervyn Barnes of the Oxford Physics Practical Course deserves special mention for running his cryostats to provide some of the illustrative data in Chapters 8 and 9. Finally, I should like to thank Chuck Mielke (Los Alamos), who provided the cover picture and data.

Lastly, I am profoundly grateful to my wife Claire, and children, Louisa and Joe, for their support and love during the preparation of this book.

J. S.
Los Alamos

February 2001

---

[7] Please email corrections to j.singleton1@physics.ox.ac.uk.

To the late
Thomas William Baverstock
Vicar of Pilling
whose enthusiasm for quantum mechanics and other wonders of creation
kindled my desire to study Physics

*The works of the Lord are great,
sought out of all them that have pleasure therein.* (Psalm 111)

Contents

1 Metals: the Drude and Sommerfeld models   **1**
  1.1 Introduction   1
  1.2 What do we know about metals?   1
  1.3 The Drude model   2
    1.3.1 Assumptions   2
    1.3.2 The relaxation-time approximation   3
  1.4 The failure of the Drude model   4
    1.4.1 Electronic heat capacity   4
    1.4.2 Thermal conductivity and the Wiedemann–Franz ratio   4
    1.4.3 Hall effect   6
    1.4.4 Summary   7
  1.5 The Sommerfeld model   7
    1.5.1 The introduction of quantum mechanics   7
    1.5.2 The Fermi–Dirac distribution function   9
    1.5.3 The electronic density of states   9
    1.5.4 The electronic density of states at $E \approx E_\mathrm{F}$ 10
    1.5.5 The electronic heat capacity   11
  1.6 Successes and failures of the Sommerfeld model   13

2 The quantum mechanics of particles in a periodic potential: Bloch's theorem   **16**
  2.1 Introduction and health warning   16
  2.2 Introducing the periodic potential   16
  2.3 Born–von Karman boundary conditions   17
  2.4 The Schrödinger equation in a periodic potential   18
  2.5 Bloch's theorem   19
  2.6 Electronic bandstructure   20

3 The nearly-free electron model   **23**
  3.1 Introduction   23
  3.2 Vanishing potential   23
    3.2.1 Single electron energy state   23
    3.2.2 Several degenerate energy levels   24
    3.2.3 Two degenerate free-electron levels   24
  3.3 Consequences of the nearly-free-electron model   26
    3.3.1 The alkali metals   27
    3.3.2 Elements with even numbers of valence electrons   27
    3.3.3 More complex Fermi surface shapes   29

4	**The tight-binding model**		**32**
	4.1 Introduction		32
	4.2 Band arising from a single electronic level		32
		4.2.1 Electronic wavefunctions	32
		4.2.2 Simple crystal structure.	33
		4.2.3 The potential and Hamiltonian	33
	4.3 General points about the formation of tight-binding bands		35
		4.3.1 The group IA and IIA metals; the tight-binding model viewpoint	36
		4.3.2 The Group IV elements	36
		4.3.3 The transition metals	37
5	**Some general points about bandstructure**		**41**
	5.1 Comparison of tight-binding and nearly-free-electron bandstructure		41
	5.2 The importance of **k**		42
		5.2.1 $\hbar\mathbf{k}$ is *not* the momentum	42
		5.2.2 Group velocity	42
		5.2.3 The effective mass	42
		5.2.4 The effective mass and the density of states	43
		5.2.5 Summary of the properties of **k**	44
		5.2.6 Scattering in the Bloch approach	45
	5.3 Holes		45
	5.4 Postscript		46
6	**Semiconductors and Insulators**		**49**
	6.1 Introduction		49
	6.2 Bandstructure of Si and Ge		50
		6.2.1 General points	50
		6.2.2 Heavy and light holes	51
		6.2.3 Optical absorption	51
		6.2.4 Constant energy surfaces in the conduction bands of Si and Ge	52
	6.3 Bandstructure of the direct-gap III–V and II–VI semiconductors		53
		6.3.1 Introduction	53
		6.3.2 General points	53
		6.3.3 Optical absorption and excitons	54
		6.3.4 Excitons	55
		6.3.5 Constant energy surfaces in direct-gap III–V semiconductors	56
	6.4 Thermal population of bands in semiconductors		56
		6.4.1 The law of mass action	56
		6.4.2 The motion of the chemical potential	58
		6.4.3 Intrinsic carrier density	58
		6.4.4 Impurities and extrinsic carriers	59
		6.4.5 Extrinsic carrier density	60
		6.4.6 Degenerate semiconductors	62

	6.4.7	Impurity bands	62
	6.4.8	Is it a semiconductor or an insulator?	62
	6.4.9	A note on photoconductivity	63

7 Bandstructure engineering — 65

- 7.1 Introduction — 65
- 7.2 Semiconductor alloys — 65
- 7.3 Artificial structures — 66
  - 7.3.1 Growth of semiconductor multilayers — 66
  - 7.3.2 Substrate and buffer layer — 68
  - 7.3.3 Quantum wells — 68
  - 7.3.4 Optical properties of quantum wells — 69
  - 7.3.5 Use of quantum wells in opto-electronics — 70
  - 7.3.6 Superlattices — 71
  - 7.3.7 Type I and type II superlattices — 71
  - 7.3.8 Heterojunctions and modulation doping — 73
  - 7.3.9 The envelope-function approximation — 74
- 7.4 Band engineering using organic molecules — 75
  - 7.4.1 Introduction — 75
  - 7.4.2 Molecular building blocks — 75
  - 7.4.3 Typical Fermi surfaces — 77
  - 7.4.4 A note on the effective dimensionality of Fermi-surface sections — 78
- 7.5 Layered conducting oxides — 78
- 7.6 The Peierls transition — 81

8 Measurement of bandstructure — 85

- 8.1 Introduction — 85
- 8.2 Lorentz force and orbits — 85
  - 8.2.1 General considerations — 85
  - 8.2.2 The cyclotron frequency — 85
  - 8.2.3 Orbits on a Fermi surface — 87
- 8.3 The introduction of quantum mechanics — 87
  - 8.3.1 Landau levels — 87
  - 8.3.2 Application of Bohr's correspondence principle to arbitrarily-shaped Fermi surfaces in a magnetic field — 89
  - 8.3.3 Quantisation of the orbit area — 90
  - 8.3.4 The electronic density of states in a magnetic field — 91
- 8.4 Quantum oscillatory phenomena — 91
  - 8.4.1 Types of quantum oscillation — 93
  - 8.4.2 The de Haas–van Alphen effect — 94
  - 8.4.3 Other parameters which can be deduced from quantum oscillations — 96
  - 8.4.4 Magnetic breakdown — 97
- 8.5 Cyclotron resonance — 97
  - 8.5.1 Cyclotron resonance in metals — 98
  - 8.5.2 Cyclotron resonance in semiconductors — 98
- 8.6 Interband magneto-optics in semiconductors — 100

	8.7	Other techniques	102
		8.7.1 Angle-resolved photoelectron spectroscopy (ARPES)	103
		8.7.2 Electroreflectance spectroscopy	104
	8.8	Some case studies	105
		8.8.1 Copper	105
		8.8.2 Recent controversy: Sr_2RuO_4	106
		8.8.3 Studies of the Fermi surface of an organic molecular metal	106
	8.9	Quasiparticles: interactions between electrons	112

9 Transport of heat and electricity in metals and semiconductors — 117
- 9.1 A brief digression; life without scattering would be difficult! — 117
- 9.2 Thermal and electrical conductivity of metals — 119
  - 9.2.1 Metals: the 'Kinetic theory' of electron transport — 119
  - 9.2.2 What do τ_σ and τ_κ represent? — 120
  - 9.2.3 Matthiessen's rule — 122
  - 9.2.4 Emission and absorption of phonons — 122
  - 9.2.5 What is the characteristic energy of the phonons involved? — 123
  - 9.2.6 Electron–phonon scattering at room temperature — 123
  - 9.2.7 Electron–phonon scattering at $T \ll \theta_D$ — 123
  - 9.2.8 Departures from the low temperature $\sigma \propto T^{-5}$ dependence — 124
  - 9.2.9 Very low temperatures and/or very dirty metals — 124
  - 9.2.10 Summary — 125
  - 9.2.11 Electron–electron scattering — 125
- 9.3 Electrical conductivity of semiconductors — 127
  - 9.3.1 Temperature dependence of the carrier densities — 127
  - 9.3.2 The temperature dependence of the mobility — 128
- 9.4 Disordered systems and hopping conduction — 129
  - 9.4.1 Thermally-activated hopping — 129
  - 9.4.2 Variable range hopping — 130

10 Magnetoresistance in three-dimensional systems — 133
- 10.1 Introduction — 133
- 10.2 Hall effect with more than one type of carrier — 133
  - 10.2.1 General considerations — 133
  - 10.2.2 Hall effect in the presence of electrons and holes — 135
  - 10.2.3 A clue about the origins of magnetoresistance — 135
- 10.3 Magnetoresistance in metals — 135
  - 10.3.1 The absence of magnetoresistance in the Sommerfeld model of metals — 135
  - 10.3.2 The presence of magnetoresistance in real metals — 137
  - 10.3.3 The use of magnetoresistance in finding the Fermi-surface shape — 138
- 10.4 The magnetophonon effect — 139

11	**Magnetoresistance in two-dimensional systems and the quantum Hall effect**	**143**
	11.1 Introduction: two-dimensional systems	143
	11.2 Two-dimensional Landau-level density of states	144
	11.2.1 Resistivity and conductivity tensors for a two-dimensional system	145
	11.3 Quantisation of the Hall resistivity	147
	11.3.1 Localised and extended states	148
	11.3.2 A further refinement– spin splitting	148
	11.4 Summary	149
	11.5 The fractional quantum Hall effect	150
	11.6 More than one subband populated	151
12	**Inhomogeneous and hot carrier distributions in semiconductors**	**154**
	12.1 Introduction: inhomogeneous carrier distributions	154
	12.1.1 The excitation of minority carriers	154
	12.1.2 Recombination	155
	12.1.3 Diffusion and recombination	155
	12.2 Drift, diffusion and the Einstein equations	156
	12.2.1 Characterisation of minority carriers; the Shockley–Haynes experiment	156
	12.3 Hot carrier effects and ballistic transport	158
	12.3.1 Drift velocity saturation and the Gunn effect	158
	12.3.2 Avalanching	160
	12.3.3 A simple resonant tunnelling structure	160
	12.3.4 Ballistic transport and the quantum point contact	161
A	**Useful terminology in condensed matter physics**	**165**
	A.1 Introduction	165
	A.2 Crystal	165
	A.3 Lattice	165
	A.4 Basis	165
	A.5 Physical properties of crystals	166
	A.6 Unit cell	166
	A.7 Wigner–Seitz cell	167
	A.8 Designation of directions	167
	A.9 Designation of planes; Miller indices	168
	A.10 Conventional or primitive?	169
	A.11 The 14 Bravais lattices	171
B	**Derivation of density of states in k-space**	**172**
	B.1 Introduction	172
	B.1.1 Density of states	173
	B.1.2 Reading	174
C	**Derivation of distribution functions**	**175**
	C.1 Introduction	175
	C.1.1 Bosons	178
	C.1.2 Fermions	178
	C.1.3 The Maxwell–Boltzmann distribution function	178
	C.1.4 Mean energy and heat capacity of the classical gas	179

D Phonons — 181
- D.1 Introduction — 181
- D.2 A simple model — 182
  - D.2.1 Extension to three dimensions — 183
- D.3 The Debye model — 185
  - D.3.1 Phonon number — 187
  - D.3.2 Summary; the Debye temperature as a useful energy scale in solids — 188
  - D.3.3 A note on the effect of dimensionality — 188

E The Bohr model of hydrogen — 191
- E.1 Introduction — 191
- E.2 Hydrogenic impurities — 192
- E.3 Excitons — 192

F Experimental considerations in measuring resistivity and Hall effect — 194
- F.1 Introduction — 194
- F.2 The four-wire method — 194
- F.3 Sample geometries — 196
- F.4 The van der Pauw method. — 197
- F.5 Mobility spectrum analysis — 198
- F.6 The resistivity of layered samples — 198

G Canonical momentum — 200

H Superconductivity — 201
- H.1 Introduction — 201
- H.2 Pairing — 201
- H.3 Pairing and the Meissner effect — 203

I List of selected symbols — 205

J Solutions and additional hints for selected exercises — 209

Index — 217

Metals: the Drude and Sommerfeld models

1.1	Introduction	1
1.2	What do we know about metals?	1
1.3	The Drude model	2
1.4	The failure of the Drude model	4
1.5	The Sommerfeld model	7
1.6	Successes and failures of the Sommerfeld model	13

1.1 Introduction

The discovery of the electron just over a century ago was a turning point in the study of condensed matter physics.[1] The existence of these tiny, charge-carrying particles in solids gave an obvious mechanism for electrical conductivity. The question then arose 'if all solids contain electrons, why do some conduct electricity whilst others do not?' We shall follow the path taken by many of the early scientists in this field and first examine metals, the class of solids in which the presence of the charge-carrying electrons is most obvious. By discovering why and how metals exhibit high electrical conductivity, we can, with luck, start to understand why other materials (glass, diamond, wood) do not.

[1] A synopsis of the early history of this field is given in the introduction of *The theory of the properties of metals and alloys*, by N.F. Mott and H. Jones (Oxford University Press, second edition 1958), an old book, but still a classic. The review includes listings of the original publications of Drude and Sommerfeld.

1.2 What do we know about metals?

In our attempt to understand metals, it is useful to list some of their properties, and contrast them with other classes of solids.

(1) The metallic state is favoured by elements; more than two thirds are metals.
(2) Metals tend to come from the left-hand side of the periodic table; therefore, a metal atom will consist of a rather tightly bound 'noble-gas-like' ionic core surrounded by a small number of more loosely-bound valence electrons.
(3) Metals form in crystal structures which have relatively large numbers n_{nn} of nearest neighbours, e.g.
   - hexagonal close-packed $n_{nn} = 12$
   - face-centred cubic $n_{nn} = 12$
   - body-centred cubic $n_{nn} = 8$ at distance a (the nearest-neighbour distance) with another 6 at $1.15\,a$.
   - These figures may be compared with typical ionic and covalent systems in which $n_{nn} \sim 4$–6).
   - The large coordination numbers and the small numbers of valence electrons (~ 1) in metals imply that the outer electrons of the metal atoms are occupying space between the ionic cores rather uniformly.
(4) The fact that the valence electrons are rather evenly spread between the ionic cores suggests that the bonds that bind the metal together are rather undirectional; this is supported by the malleability of metals. By

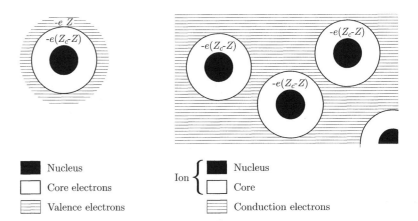

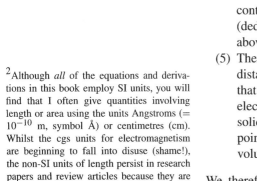

Fig. 1.1 Schematic representations of a single, isolated metal atom and a solid metal. The atom (left) consists of a nucleus (size greatly exaggerated!) of charge $+eZ_c$ surrounded by the core electrons, which provide charge $-e(Z_c - Z)$, and Z valence electrons, of charge $-eZ$. In the solid metal (right), the core electrons remain bound to the nucleus, but the valence electrons move throughout the solid, becoming conduction electrons.

[2] Although *all* of the equations and derivations in this book employ SI units, you will find that I often give quantities involving length or area using the units Angstroms (= 10^{-10} m, symbol Å) or centimetres (cm). Whilst the cgs units for electromagnetism are beginning to fall into disuse (shame!), the non-SI units of length persist in research papers and review articles because they are so very convenient (e.g. atomic or ionic spacings in solids fall in the range 2–3 Å, which somehow seems easier to remember than 0.2–0.3 nm). It is therefore necessary to get used to these units, and this is the reason that I have not excised them from the text.

[3] In the subsequent sections, the Drude model will be seen to fail rather spectacularly in some respects. This does not lessen my great respect for Drude; after all, his original paper was published less than three years after the discovery of the electron!

contrast, ionic and covalent solids, with their more directional bonds (deduced from the smaller number of nearest neighbours mentioned above) are brittle.

(5) There is a lot of 'empty space' in metals; e.g. Li has an interatomic distance of 3 Å, but the ionic radius is only about 0.5 Å.[2] This implies that there is a great deal of volume available for the valence (conduction) electrons to move around in. This is actually one of the reasons why solid metals are stable; the valence electrons have a much lower zero-point energy when they can spread themselves out through these large volumes than when they are confined to one atom.

We therefore picture the metal as an array of widely spaced, small ionic cores, with the mobile valence electrons spread through the volume between (see Fig. 1.1). The positive charges of the ionic cores act to provide charge neutrality for the valence electrons, confining the electrons within the solid (see Fig. 1.2).

1.3 The Drude model

1.3.1 Assumptions

The Drude model was the first attempt to use the idea of a 'gas' of electrons, free to move between positively charged ionic cores, in order to explain the properties of metals.[3] The assumptions of the Drude model are

- a *collision* indicates the scattering of an electron by (and only by) an ionic core, i.e. the electrons do not 'collide' with anything else;
- between collisions, electrons do not interact with each other (the **independent electron approximation**) or with ions (the **free electron approximation**);
- collisions are instantaneous and result in a change in electron velocity;
- an electron suffers a collision with probability per unit time τ^{-1} (the **relaxation-time approximation**), i.e. τ^{-1} is the scattering rate;
- electrons achieve thermal equilibrium with their surroundings only through collisions.

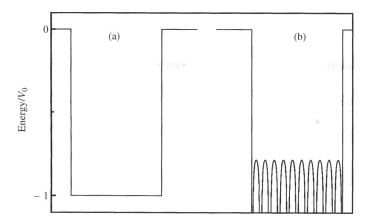

Fig. 1.2 Schematic of the electronic potential assumed in the Drude and Sommerfeld models of metals (a). The small-scale details of the potential due to the ionic cores (shown schematically in (b)) are replaced by an average potential $-V_0$. In such a picture, the ionic cores merely maintain charge neutrality, hence keeping the electrons within the metal; the metal sample acts as a 'box' containing electrons which are free to move within it.

These assumptions should be contrasted with those of the kinetic theory of conventional gases.[4] For example, in conventional kinetic theory, equilibrium is achieved by collisions *between* gas molecules; in the Drude model, the 'molecules' (electrons) do not interact with each other at all!

Before discussing the results of the Drude model in more detail, we shall examine the way in which the scattering rate can be included in the equations of motion of the electrons. This approach will also prove to be very useful in more sophisticated treatments than the Drude model.

[4] See e.g. *Three phases of matter*, by Alan J. Walton, second edition (Clarendon Press, Oxford, 1983) Chapters 5–7.

1.3.2 The relaxation-time approximation

The current density **J** due to electrons, of average velocity **v**, is[5]

$$\mathbf{J} = -ne\mathbf{v} = -\frac{ne}{m_e}\mathbf{p}, \quad (1.1)$$

[5] I shall use the symbol e to denote the *magnitude* of the electronic charge, i.e. $-e$ is the charge of the electron, $+e$ is the charge on the positron.

where m_e is the electron mass, n is the electron density and **p** is the average electron momentum. Consider the evolution of **p** in time δt under the action of an external force $\mathbf{f}(t)$. The probability of a collision during δt is $\frac{\delta t}{\tau}$, i.e. the probability of surviving without colliding is $(1 - \frac{\delta t}{\tau})$. For electrons that don't collide, the increase in momentum $\delta\mathbf{p}$ is given by[6]

$$\delta\mathbf{p} = \mathbf{f}(t)\delta t + O(\delta t)^2. \quad (1.2)$$

[6] The $O(\delta t)^2$ comes from the fact that I have allowed the force to vary with time. In other words, the force at time $t + \delta t$ will be slightly different than that at time t; e.g. if δt is very small, then $\mathbf{f}(t + \delta t) \approx \mathbf{f}(t) + (\mathrm{d}\mathbf{f}/\mathrm{d}t)\delta t$.

Therefore, the contribution to the average electron momentum from the electrons that do not collide is

$$\mathbf{p}(t + \delta t) = (1 - \frac{\delta t}{\tau})(\mathbf{p}(t) + \mathbf{f}(t)\delta t + O(\delta t)^2). \quad (1.3)$$

Note that the contribution to the average momentum from electrons which *have* collided will be of order $(\delta t)^2$ and therefore negligible. This is because they constitute a fraction $\sim \delta t/\tau$ of the electrons and because the momentum that they will have acquired since colliding (each collision effectively randomises their momentum) will be $\sim \mathbf{f}(t)\delta t$.

Equation 1.3 can then be rearranged to give, in the limit $\delta t \to 0$

$$\frac{\mathrm{d}\mathbf{p}(t)}{\mathrm{d}t} = -\frac{\mathbf{p}(t)}{\tau} + \mathbf{f}(t). \quad (1.4)$$

In other words, the collisions produce a frictional damping term. This idea will have many applications, even beyond the Drude model; we shall now use it to derive the electrical conductivity.

The electrical conductivity σ is defined by

$$\mathbf{J} = \sigma \mathbf{E}, \tag{1.5}$$

where $\mathbf{E}$ is the electric field. To find σ, we substitute $\mathbf{f} = -e\mathbf{E}$ into eqn 1.4; we also set $\frac{d\mathbf{p}(t)}{dt} = 0$, as we are looking for a steady-state solution. The momentum can then be substituted into eqn 1.1, to give

$$\sigma = ne^2\tau/m_e. \tag{1.6}$$

If we substitute the room-temperature value of σ for a typical metal[7] along with a typical $n \sim 10^{22}$–10^{23} cm^{-3} into this equation, a value of $\tau \sim$ 1–10 fs emerges. In Drude's picture, the electrons are the particles of a classical gas, so that they will possess a mean kinetic energy

$$\frac{1}{2} m_e \langle v^2 \rangle = \frac{3}{2} k_B T \tag{1.7}$$

(see Section C.1.4; the brackets $\langle \rangle$ denote the mean value of a quantity). Using this expression to derive a typical classical room temperature electron speed, we arrive at a mean free path $v\tau \sim$ 0.1–1 nm. This is roughly the same as the interatomic distances in metals, a result consistent with the Drude picture of electrons colliding with the ionic cores. However, we shall see later that the Drude model very seriously underestimates typical electronic velocities.

[7] Most metals have resistivities ($1/\sigma$) in the range $\sim 1 - 20$ $\mu\Omega$cm at room temperature. Some typical values are tabulated on page 8 of *Solid state physics*, by N.W Ashcroft and N.D. Mermin (Holt, Rinehart and Winston, New York, 1976).

1.4 The failure of the Drude model

1.4.1 Electronic heat capacity

The Drude model predicts the electronic heat capacity to be the classical 'equipartition of energy' result[8] i.e.

$$C_{el} = \frac{3}{2} n k_B. \tag{1.8}$$

This is independent of temperature. Experimentally, the low-temperature heat capacity of metals follows the relationship (see Figs. 1.3 and 1.4)

$$C_V = \gamma T + \alpha T^3. \tag{1.9}$$

The second term is obviously the phonon (Debye) component (see Section D.3), leading us to suspect that $C_{el} = \gamma T$. Indeed, even at room temperature, the electronic component of the heat capacity of metals is much smaller than the Drude prediction. This is obviously a severe failing of the model.

[8] See Section C.1.4 and/or any statistical mechanics book (e.g. *Three phases of matter*, by Alan J. Walton, second edition (Clarendon Press, Oxford, 1983) page 126; *Statistical physics*, by Tony Guenault (Routledge, London, 1988) Section 3.2.2.

1.4.2 Thermal conductivity and the Wiedemann–Franz ratio

The thermal conductivity κ is defined by the equation

$$\mathbf{J}_q = -\kappa \nabla T, \tag{1.10}$$

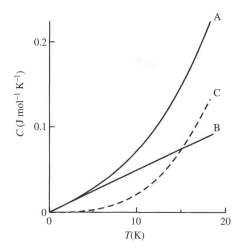

Fig. 1.3 Heat capacity of Co at low temperatures. A is the experimental curve, B is the electronic contribution γT and C is the Debye component AT^3. (Data from G. Duyckaerts, *Physica* **6**, 817 (1939).)

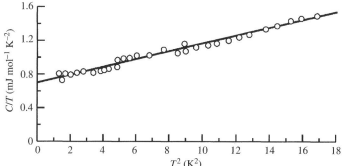

Fig. 1.4 Plot of C_V/T versus T^2 for Cu; the intercept gives γ. (Data from W.S. Corak *et al.*, *Phys. Rev.* **98**, 1699 (1955).)

where $\mathbf{J}_q$ is the flux of heat (i.e. energy per second per unit area). The Drude model assumes that the conduction of heat in metals is almost entirely due to electrons, and uses the kinetic theory expression[9] for κ, i.e.

$$\kappa = \frac{1}{3}\langle v^2 \rangle \tau C_{\text{el}}. \quad (1.11)$$

C_{el} is taken from eqn 1.8 and the speed takes the classical thermal value, i.e. $\frac{1}{2}m_e \langle v^2 \rangle = \frac{3}{2}k_B T$ (see Section C.1.4).

The fortuitous success of this approach came with the prediction of the Wiedemann–Franz ratio, κ/σ. It had been found that the Wiedemann–Franz ratio divided by the temperature, $L = \kappa/(\sigma T)$,[10] was close to the constant value $\sim 2.5 \times 10^{-8} \text{W}\Omega\text{K}^{-2}$ for many metals at room temperature.[11] Substituting eqns 1.6 and 1.11 into this ratio leads to

$$\frac{\kappa}{\sigma T} = \frac{3}{2}\frac{k_B^2}{e^2} \approx 1.1 \times 10^{-8} \text{ W}\Omega\text{K}^{-2}. \quad (1.12)$$

However, in spite of this apparent success, the individual components of the model are very wrong: e.g. C_{el} in the Drude model is at least two orders of magnitude bigger than the experimental values at room temperature! Furthermore, experimentally $\kappa/(\sigma T)$ drops away from its constant value at temperatures below room temperature (see Fig. 1.5); the Drude model cannot explain this behaviour.

[9] See any kinetic theory book, e.g. *Three phases of matter*, by Alan J. Walton, second edition (Clarendon Press, Oxford, 1983) Section 7.3.

[10] $L = \kappa/(\sigma T)$ is often referred to as the Lorenz number.

[11] Typical values of κ and σ for metals are available in *Tables of physical and chemical constants*, by G.W.C. Kaye and T.H. Laby (Longmans, London, 1966); find the book and check out the Lorenz number for yourself!

6 Metals: the Drude and Sommerfeld models

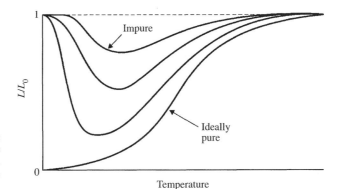

Fig. 1.5 Schematic of experimental variation of $L = \kappa/(\sigma T)$ with temperature; L_0 is the theoretical value of L derived in Chapter 9. The high temperature limit of the figure represents room temperature.

1.4.3 Hall effect

The Hall effect occurs when a magnetic field **B** is applied perpendicular to a current density **J** flowing in a sample (see Fig. 10.1 for the geometry of a typical experiment). A **Hall voltage** is developed in the direction perpendicular to both **J** and **B**. For simplicity, we shall assume that the current flows parallel to the x direction and that the magnetic field is parallel to to z direction; the Hall voltage will then be developed in the y direction.

The presence of both electric and magnetic fields means that the force on an electron is now $\mathbf{f} = -e\mathbf{E} - e\mathbf{v} \times \mathbf{B}$. Equation 1.4 becomes

$$\frac{d\mathbf{v}}{dt} = -\frac{e}{m_e}\mathbf{E} - \frac{e}{m_e}\mathbf{v} \times \mathbf{B} - \frac{\mathbf{v}}{\tau}. \tag{1.13}$$

In steady state, the left-hand side vanishes, and the two components of the equation read

$$v_x = -\frac{e\tau}{m_e}E_x - \omega_c\tau v_y \tag{1.14}$$

and

$$v_y = -\frac{e\tau}{m_e}E_y + \omega_c\tau v_x, \tag{1.15}$$

where we have written $\omega_c = eB/m_e$ (ω_c is of course the classical cyclotron angular frequency). If we impose the condition $v_y = 0$ (no current in the y direction; see Fig. 10.1) we have

$$\frac{E_y}{E_x} = -\omega_c\tau. \tag{1.16}$$

Writing $J_x = -env_x$, eqns 1.14 and 1.16 can be combined to give

$$R_H \equiv \frac{E_y}{J_x B} = -\frac{1}{ne}. \tag{1.17}$$

R_H is known as the **Hall coefficient**.

The Drude model therefore predicts the surprising fact that R_H is independent of both the magnitude of B and the scattering time τ. In reality, however, R_H is found to vary with magnetic field (see Fig. 1.6) In many cases (see Table 1.1) the Hall coefficient is even *positive*! I shall offer an explanation of these effects in Chapter 10, where the Hall effect will be treated much more rigorously.

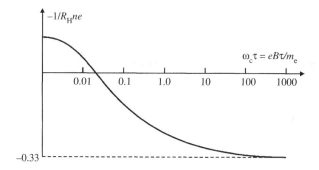

Fig. 1.6 $-1/(R_H ne)$ for aluminium as a function of magnetic field. Data taken from R. Lück, *Phys. Stat. Sol.* **18**, 49 (1966).

Table 1.1 Observed and calculated values of the Hall coefficient R_H in units of $10^{-11} \text{m}^3 \text{C}^{-1}$ for several metals. The calculated values assume j free electrons per atom.

Metal	Observed R_H	Calculated R_H	j
Li	−17.0	−13.1	1
Na	−25.0	−24.4	1
Cu	−5.5	−7.4	1
Ag	−8.4	−10.4	1
Zn	+4.1	−4.6	2
Cd	+6.0	−6.5	2

1.4.4 Summary

Although the Drude model was a brave attempt to apply classical kinetic theory to the electrons in a metal, it is plainly flawed in many respects.[12]

In order to start to remedy the most severe failings (e.g. the electronic heat capacity, the thermal conductivity) we must treat the electrons as quantum-mechanical particles (fermions), rather than the molecules of a classical gas. As we shall see in the next section, this will dramatically constrain the way in which electrons can involve themselves in the thermal properties of the metal. Our treatment is based on the work of Sommerfeld,[13] who introduced quantum statistics by replacing the Maxwell–Boltzmann distribution of the Drude model by the Fermi–Dirac distribution function.[14]

1.5 The Sommerfeld model

1.5.1 The introduction of quantum mechanics

Quantum mechanics tells us that we can only have a certain number of states N per unit volume of phase space, where

$$N = \left(\frac{1}{2\pi}\right)^j V_{kj} V_{rj}, \qquad (1.18)$$

where j is the number of dimensions, V_{kj} is the k-space volume and V_{rj} is the r-space (i.e. real) volume (see Appendix B for a derivation). For example, in

[12] I re-emphasise my respect for the pioneer Drude; the model was *the first attempt* to understand metals on a microscopic level. Although flawed, its prediction of the correct order of magnitude for the Wiedemann–Franz ratio provoked a large amount of further work which led to the Sommerfeld model.

[13] For citations of the original papers, see *The theory of the properties of metals and alloys*, by N.F. Mott and H. Jones (Oxford University Press, second edition 1958).

[14] The assumptions used in deriving the Maxwell–Boltzmann and Fermi–Dirac distribution functions are treated in Sections C.1.2 and C.1.3.

three dimensions

$$N = \left(\frac{1}{2\pi}\right)^3 V_{k3} V_{r3}. \tag{1.19}$$

Now we assume electrons free to move within the solid; once again we ignore the details of the periodic potential due to the ionic cores (see Fig. 1.2(b)),[15] replacing it with a mean potential $-V_0$ (see Fig. 1.2(a)). The energy of the electrons is then

$$E(\mathbf{k}) = -V_0 + \frac{p^2}{2m_e} = -V_0 + \frac{\hbar^2 k^2}{2m_e}. \tag{1.20}$$

However, we are at liberty to set the origin of energy; for convenience we choose $E = 0$ to correspond to the average potential within the metal (Fig. 1.2(a)), so that

$$E(\mathbf{k}) = \frac{\hbar^2 k^2}{2m_e}. \tag{1.21}$$

Let us now imagine gradually putting free electrons into a metal. At $T = 0$, the first electrons will occupy the lowest energy (i.e. lowest $|\mathbf{k}|$) states; subsequent electrons will be forced to occupy higher and higher energy states (because of Pauli's exclusion principle). Eventually, when all N electrons are accommodated, we will have filled a sphere of k-space, with

$$N = 2\left(\frac{1}{2\pi}\right)^3 \frac{4}{3}\pi k_F^3 V_{r3}, \tag{1.22}$$

where we have included a factor 2 to cope with the electrons' spin degeneracy and where k_F is the k-space radius (the *Fermi wavevector*) of the sphere of filled states. Rearranging, remembering that $n = N/V_{r3}$, gives

$$k_F = (3\pi^2 n)^{\frac{1}{3}}. \tag{1.23}$$

The corresponding electron energy is

$$E_F = \frac{\hbar^2 k_F^2}{2m_e} = \frac{\hbar^2}{2m_e}(3\pi^2 n)^{\frac{2}{3}}. \tag{1.24}$$

This energy is called the *Fermi energy*.

The substitution of typical metallic carrier (i.e. electron) densities into eqn 1.24,[16] produces values of E_F in the range ~ 1.5–15 eV (i.e. $\sim$ atomic energies) or $E_F/k_B \sim 20 - 100 \times 10^3$ K (i.e. $\gg$ room temperature). Typical Fermi wavevectors are $k_F \sim 1/$(the atomic spacing) $\sim$ the Brillouin zone size (see Chapter 2) $\sim$ typical X-ray k-vectors. The velocities of electrons at the Fermi surface are $v_F = \hbar k_F/m_e \sim 0.01c$. Although the Fermi surface is the $T = 0$ ground-state of the electron system, the electrons present are enormously energetic!

The presence of the Fermi surface has many consequences for the properties of metals; e.g. the main contribution to the bulk modulus of metals comes from the Fermi 'gas' of electrons.[17]

---

[15] Many condensed matter physics texts spend time on hand-waving arguments which supposedly justify the fact that electrons do not interact much with the ionic cores. Justifications used include 'the ionic cores are very small'; 'the electron–ion interactions are strongest at small separations but the Pauli exclusion principle prevents electrons from entering this region'; 'screening of ionic charge by the mobile valence electrons'; 'electrons have higher average kinetic energies whilst traversing the deep ionic potential wells and therefore spend less time there'. We shall see in Chapters 2–5 that it is the *translational* symmetry of the potential due to the ions which allows the electrons to travel for long distances without scattering from anything.

[16] Metallic carrier densities are usually in the range $\sim 10^{22}$–10^{23} cm^{-3}. Values for several metals are tabulated in *Solid state physics*, by N.W Ashcroft and N.D. Mermin (Holt, Rinehart and Winston, New York, 1976), page 5. See also Exercise 1.1.

[17] See Exercise 1.2; selected bulk moduli for metals have been tabulated on page 39 of *Solid state physics*, by N.W Ashcroft and N.D. Mermin (Holt, Rinehart and Winston, New York, 1976).

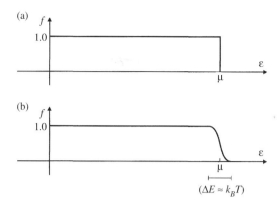

Fig. 1.7 The Fermi–Dirac distribution function at $T = 0$ and at a finite temperature $T \ll E_F/k_B$.

1.5.2 The Fermi–Dirac distribution function

In order to treat the thermal properties of electrons at higher temperatures, we need an appropriate distribution function for electrons from statistical mechanics. The Fermi–Dirac distribution function is (see Appendix C)

$$f_D(E, T) = \frac{1}{e^{(E-\mu)/k_B T} + 1}, \quad (1.25)$$

where μ is the **chemical potential**; f_D gives the probability of occupation of a state of energy E. One of the definitions of the chemical potential[18] is that it is the energy at which the probability of occupation is $\frac{1}{2}$ (see also Section C.1).

Figure 1.7 shows f_D at $T = 0$ and at a finite temperature $T \ll E_F/k_B$ (e.g. room temperature). Considering first the $T = 0$ figure, we see that

$$E_F \equiv \mu(T = 0). \quad (1.26)$$

This defines E_F.

Turning to the finite T figure, f_D only varies significantly within $\sim k_B T$ of μ. This has two implications.

(1) As $k_B T \ll E_F$, this implies that $\mu \approx E_F$ (see Exercise 5). In much of what follows, we shall make use of Fermi–Dirac statistics. Strictly the chemical potential μ is the fundamental energy used in the Fermi–Dirac distribution function f_D. However, for a typical metal at all accessible temperatures (i.e. until it melts) $\mu \approx E_F \equiv \mu(T = 0)$. The substitution $\mu \to E_F$ made in the thermodynamic analysis below (for convenience) results in minimal errors.

(2) Only electrons with energies within $\sim k_B T$ of μ, i.e. E_F, will be able to contribute to thermal processes, transport etc. Electrons further below μ will be unable to acquire sufficient thermal energy to be excited into empty states; states more than $\sim k_B T$ above μ will be empty.

[18]The chemical potential is often (confusingly) referred to as the **Fermi level**. Although you will see this usage in other books, I have tried to stick to *chemical potential*, to avoid confusion with *the Fermi energy*. Furthermore, I find the term 'Fermi level' defective, in that it implies that there is a quantum-mechanical state (level) present at that energy. As we shall see when we come to treat semiconductors (Chapter 6), there need be no such state at that energy.

1.5.3 The electronic density of states

In subsequent derivations, it will be useful to have a function that describes the number of electron states in a particular energy range. We start by substituting

10 Metals: the Drude and Sommerfeld models

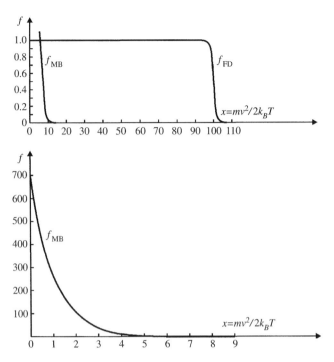

Fig. 1.8 Comparison of the Maxwell–Boltzmann and Fermi–Dirac distribution functions using parameters relevant for a typical metal at room temperature. The concentration of filled states at low energies in the Maxwell–Boltzmann distribution is the reason why the Drude model fails to describe the thermal properties of metals.

a general value of k in eqn 1.22 instead of k_F. This gives $n(\mathbf{k})$, the number of states per unit volume of r-space with wavevectors less than $|\mathbf{k}|$ (see also Section B.1.1). For free electrons, with energy $E = \hbar^2 \mathbf{k}^2 / 2m_e$, one can then define a density of states, $g(E)$, where $g(E)\mathrm{d}E$ is the number of electrons per unit volume of r-space with energies between E and $E + \mathrm{d}E$:

$$g(E) \equiv \frac{\mathrm{d}n}{\mathrm{d}E} = \frac{\mathrm{d}n}{\mathrm{d}k}\frac{\mathrm{d}k}{\mathrm{d}E} = \frac{1}{2\pi^2}\left(\frac{2m_e}{\hbar^2}\right)^{\frac{3}{2}} E^{\frac{1}{2}}. \tag{1.27}$$

Before going on to use $g(E)$ in calculating the electronic heat capacity, I shall point out a very useful expression for $g(E)$ at energies close to the Fermi energy.

1.5.4 The electronic density of states at $E \approx E_F$

As only the electrons within $\sim k_B T$ of E_F are able to take part in thermal processes, only the density of electron states at the Fermi energy, $g(E_F)$, will be particularly important in many calculations. In the free-electron approximation, the Fermi energy is given by $E_F = \frac{\hbar^2}{2m_e}(3\pi^2 n)^{\frac{2}{3}}$ (see eqn 1.24). Taking natural logarithms gives

$$\ln(E_F) = \frac{2}{3}\ln(n) + \text{const.} \tag{1.28}$$

Differentiating (i.e. making infinitessimal changes in E_F and n) gives

$$\frac{\mathrm{d}E_F}{E_F} = \frac{2}{3}\frac{\mathrm{d}n}{n}. \tag{1.29}$$

We rearrange to obtain[19]

$$\frac{dn}{dE_F} \equiv g(E_F) = \frac{3}{2}\frac{n}{E_F}. \quad (1.30)$$

1.5.5 The electronic heat capacity

Let $U(T)$ be the energy of the electron system at temperature T. At absolute zero, the electronic states are filled up to $E = E_F$ and empty for higher energies. $U(0)$ can therefore be evaluated rather easily:

$$U(0) = \int_0^{E_F} E g(E) dE = \frac{E_F^{\frac{5}{2}}}{5\pi^2}\left(\frac{2m_e}{\hbar^2}\right)^{\frac{3}{2}}, \quad (1.31)$$

where $g(E)$ has been taken from eqn 1.27.

At finite temperature, electrons are excited into higher levels, and the expression for the energy of the electron system becomes slightly more complicated

$$U(T) = \int_0^{\infty} E g(E) f_D(E,T) dE$$

$$= \frac{1}{2\pi^2}\left(\frac{2m_e}{\hbar^2}\right)^{\frac{3}{2}} \int_0^{\infty} \frac{E^{\frac{3}{2}}}{(e^{(E-\mu)/k_B T} + 1)} dE. \quad (1.32)$$

The integral in eqn 1.32 is a member of the family of so-called **Fermi–Dirac integrals**

$$F_j(y_0) = \int_0^{\infty} \frac{y^j}{e^{(y-y_0)} + 1} dy. \quad (1.33)$$

There are no general analytical expressions for such integrals, but certain asymptotic forms exist.[20] Comparison with eqn 1.32 shows that $y_0 \equiv \mu/k_B T$; we have seen that for all practical temperatures, $\mu \approx E_F \gg k_B T$. Hence, the asymptotic form that is of interest for us is the one for which y_0 is very large and positive:

$$F_j(y_0) \approx \frac{y_0^{j+1}}{j+1}\left(1 + \frac{\pi^2 j(j+1)}{6 y_0^2} + O(y_0^{-4}) + ...\right) \quad (1.34)$$

Using eqn 1.34 to evaluate eqn 1.32 in the limit $\mu \gg k_B T$ yields

$$U(T) = \frac{2}{5}\mu^{\frac{5}{2}}\{1 + \frac{5}{8}\left(\frac{\pi k_B T}{\mu}\right)^2\}\frac{1}{2\pi^2}\left(\frac{2m_e}{\hbar^2}\right)^{\frac{3}{2}}. \quad (1.35)$$

We are left with the problem that μ is temperature dependent; μ is determined by the constraint that the total number of electrons must remain constant:

$$n = \int_0^{\infty} E_F g(E) f_D(E,T) dE$$

$$= \frac{1}{2\pi^2}\left(\frac{2m_e}{\hbar^2}\right)^{\frac{3}{2}} \int_0^{\infty} \frac{E^{\frac{1}{2}}}{(e^{(E-\mu)/k_B T} + 1)} dE. \quad (1.36)$$

---

[19] Note that this expression contains only n and E_F; the mass of the electron and the spin degeneracy (factor 2) which feature in the prefactors in eqns 1.24 and 1.27 do not show up. This means that eqn 1.30 has a rather wider applicability than its simple derivation might suggest; remember this when you come to do Exercise 5.3.

[20] The asymptotic forms were originally treated by Sommerfeld (*Z. Physik* **47**, 1 (1927); an English version is given in *Rev. Mod. Phys.* **3**, 1 (1931)). J. McDougall and E.C. Stoner have tabulated a wide range of asymptotic forms in *Phil. Trans. Roy. Soc.* **A237**, 67 (1968). Chapter 2 and Appendix C of *Solid state physics*, by N.W Ashcroft and N.D. Mermin (Holt, Rinehart and Winston, New York, 1976) show some of the algebra used in evaluating the asymptotic forms for $y_0 \gg 1$.

This contains the $j = \frac{1}{2}$ Fermi–Dirac integral, which may also be evaluated using eqn 1.34 (as $\mu \gg k_B T$; see Exercise 1.5) to yield

$$\mu \approx E_F \{1 - \frac{\pi^2}{12}\left(\frac{k_B T}{\mu}\right)^2\}. \tag{1.37}$$

Therefore, using a polynomial expansion

$$\mu^{\frac{5}{2}} \approx E_F^{\frac{5}{2}} \{1 - \frac{5\pi^2}{24}\left(\frac{k_B T}{\mu}\right)^2 .....\}, \tag{1.38}$$

Combining eqns 1.35 and 1.38 and neglecting terms of order $(k_B T/E_F)^4$ gives

$$U(T) \approx \frac{2}{5} E_F^{\frac{5}{2}} \{1 - \frac{5\pi^2}{24}\left(\frac{k_B T}{\mu}\right)^2\}\{1 + \frac{5}{8}\left(\frac{\pi k_B T}{\mu}\right)^2\} \frac{1}{2\pi^2}\left(\frac{2m_e}{\hbar^2}\right)^{\frac{3}{2}}.$$

$$\approx \frac{2}{5} E_F^{\frac{5}{2}} \{1 + \frac{5}{12}\left(\frac{\pi k_B T}{\mu}\right)^2\} \frac{1}{2\pi^2}\left(\frac{2m_e}{\hbar^2}\right)^{\frac{3}{2}}. \tag{1.39}$$

Thus far, it was essential to keep the temperature dependence of μ in the algebra in order to get the prefactor of the temperature-dependent term in eqn 1.39 correct. However, eqn 1.37 shows that at temperatures $k_B T \ll \mu$ (e.g. at room temperature in typical metal), $\mu \approx E_F$ to a high degree of accuracy. In order to get a reasonably accurate estimate for $U(T)$ we can therefore make the substitution $\mu \to E_F$ in denominator of the second term of the bracket of eqn 1.39 to give

$$U(T) = U(0) + \frac{n\pi^2 k_B^2 T^2}{4 E_F}, \tag{1.40}$$

where we have substituted in the value for $U(0)$ from eqn 1.31. Differentiating, we obtain

$$C_{el} \equiv \frac{\partial U}{\partial T} = \frac{1}{2}\pi^2 n \frac{k_B^2 T}{E_F}. \tag{1.41}$$

Equation 1.41 looks very good as it

(1) is proportional to T, as are experimental data (see Figs. 1.3 and 1.4);
(2) is a factor $\sim \frac{k_B T}{E_F}$ smaller than the classical (Drude) value, as are experimental data.

[21] The method for deriving the electronic heat capacity used in books such as *Introduction to solid state physics*, by Charles Kittel, seventh edition (Wiley, New York, 1996) looks at first sight rather simpler than the one that I have followed. However, Kittel's method contains a hidden trap which is not mentioned; it ignores $(d\mu/dT)$, which is said to be negligible. This term is actually of comparable size to all of the others in Kittel's integral (see *Solid state physics*, by N.W Ashcroft and N.D. Mermin (Holt, Rinehart and Winston, New York, 1976) Chapter 2), but happens to disappear because of a cunning change of variables. By the time all of this has explained in depth, Kittel's method looks rather more tedious and less clear, and I do not blame him for wishing to gloss over the point!

Just in case the algebra of this section has seemed like hard work,[21] note that a reasonably accurate estimate of C_{el} can be obtained using the following reasoning. At all practical temperatures, $k_B T \ll E_F$, so that we can say that only the electrons in an energy range $\sim k_B T$ on either side of $\mu \approx E_F$ will be involved in thermal processes. The number density of these electrons will be $\sim k_B T g(E_F)$. Each electron will be excited to a state $\sim k_B T$ above its ground-state ($T = 0$) energy. A reasonable estimate of the thermal energy of the system will therefore be

$$U(T) - U(0) \sim (k_B T)^2 g(E_F) = \frac{3}{2} n k_B T \left(\frac{k_B T}{E_F}\right),$$

where I have substituted the value of $g(E_F)$ from eqn 1.30. Differentiating with respect to T, we obtain

$$C_{el} = 3 n k_B \left(\frac{k_B T}{E_F}\right),$$

which is within a factor 2 of the more accurate method.

1.6 Successes and failures of the Sommerfeld model

The Sommerfeld model is a great improvement on the Drude model; it can successfully explain

- the temperature dependence and magnitude of C_{el};
- the approximate temperature dependence and magnitudes of the thermal and electrical conductivities of metals, and the Wiedemann–Franz ratio (see Chapter 9);
- the fact that the electronic magnetic susceptibility is temperature independent.[22]

It cannot explain

[22]The susceptibility of metals in the Sommerfeld model is derived in Chapter 7 of *Magnetism in condensed matter*, by S. Blundell (OUP 2001).

- the Hall coefficients of many metals (see Fig. 1.6 and Table 1.1; Sommerfeld predicts $R_H = -1/ne$);
- the magnetoresistance exhibited by metals (see Chapter 10);
- other parameters such as the thermopower;
- the shapes of the Fermi surfaces in many real metals;
- the fact that some materials are insulators and semiconductors (i.e. not metals).

Furthermore, it seems very intellectually unsatisfying to completely disregard the interactions between the electrons and the ionic cores, except as a source of instantaneous 'collisions'. This is actually the underlying source of the difficulty; to remedy the failures of the Sommerfeld model, we must reintroduce the interactions between the ionic cores and the electrons. In other words, we must introduce the periodic potential of the solid.

Further reading

A more detailed treatment of the topics in this chapter is given in *Solid state physics*, by N.W Ashcroft and N.D. Mermin (Holt, Rinehart and Winston, New York, 1976), Chapters 1–3. Other useful information can be found in *Electricity and magnetism*, by B.I. Bleaney and B. Bleaney, third edition (Oxford University Press, Oxford, 1989), Chapter 11, *Solid state physics*, by G. Burns (Academic Press, Boston, 1995), Sections 9.1–9.14, *Electrons in metals and semiconductors*, by R.G. Chambers (Chapman and Hall, London, 1990), Chapters 1 and 2, and *Introduction to solid state physics*, by Charles Kittel, seventh edition (Wiley, New York, 1996), Chapters 6 and 7.

14 *Metals: the Drude and Sommerfeld models*

Exercises

I use some of the following exercises to introduce the idea of an effective mass m^* in a gentle fashion. As we shall see in later chapters, the effective mass is a convenient way of describing how the ionic potential within a crystal modifies the dynamics of an electron. The exercises that follow contain the most simple manifestation of the effective mass;[23] we assume that the energies of the electrons are given by $E = \hbar^2 k^2/2m^*$, where m^* is a constant, rather than by the free-electron expression $E = \hbar^2 k^2/2m_e$.

In all of the following you may assume that the number of k states in V_{rj}, a j-dimensional volume of r-space, and V_{kj}, a j-dimensional volume of k-space, is given by $(\frac{1}{2\pi})^j V_{rj} V_{kj}$ (see Appendix B).

(1.1)   a) A sample contains N electrons, which behave as free particles with energy $E = \hbar^2 k^2/2m^*$, where m^* is an effective mass. Derive formulae for the Fermi energy (E_F) in the following cases.
    (i) The sample is one-dimensional (1D), and is of length d.
    (ii) The sample is two-dimensional (2D), and is of area A.
    (iii) The sample is three-dimensional (3D), and is of volume $\mathcal{V}$.
In each case, show that the density of states at E_F is of the form
$$g(E_F) = \xi \frac{n}{E_F} \quad (1.42)$$
with ξ a number ~ 1 and
$$n = \begin{cases} N/d & (1D) \\ N/A & (2D) \\ N/\mathcal{V} & (3D). \end{cases} \quad (1.43)$$
Hence show that, irrespective of the number of dimensions, $g(E_F)$ is proportional to m^*. (This observation will be important later on).

  b) Calculate E_F in both eV (electron volts) and K (kelvins) for (a) Copper, which has a fcc structure with $a = 0.361$ nm and is monovalent (assume that $m^* \approx m_e$); (b) a GaAs-(Ga,Al)As heterojunction containing 3×10^{11} cm^{-2} electrons with $m^* \approx 0.07 m_e$ (never mind what a heterojunction is; look at the carrier density to see its dimensionality)[24] and (c) a one-dimensional organic conductor whose unit cells each contribute one mobile electron ($m^* \sim m_e$) and are 8 Å long.

(1.2) **Thermodynamics of the Fermi surface.**

  a) Show that the mean energy of each electron in a 3D free electron system at $T = 0$ is $\frac{3}{5} E_F$. Hence or otherwise show that the bulk modulus B_m of such an electron system is given by
$$B_m = -\mathcal{V}\left(\frac{\partial P}{\partial \mathcal{V}}\right)_N = \frac{2}{3} n E_F, \quad (1.44)$$
where all the symbols have the same meaning as in Exercise 1.1. (Hint: remember that $dU = T dS - P d\mathcal{V}$.)

  b) Substitute values of n and E_F for copper (see Question 1) into your expression for B_m and compare with the actual bulk modulus of the metal itself. The two figures should be of the same order of magnitude, i.e. metals would probably be pools of gunge lying on the floor if they didn't have a Fermi surface.

(1.3)   a) Explain briefly why the low-temperature heat capacity of a metal is often plotted in the form C/T versus T^2.

  b) Using the Sommerfeld formalism (free electrons plus quantum statistics), derive a formula for the intercept γ of such a curve with the C/T axis at $T^2 = 0$.

  c) Copper has a fcc structure with $a = 0.361$ nm and is monovalent; estimate γ for copper, assuming that $m^* = m_e$. Compare your estimate with the experimental value 0.695 mJmol^{-1}K^{-2}. Why do you think that your answer is a little smaller than the experimental value?

  d) The metallic compound UPt$_3$ has $\gamma \approx 450$ mJmol^{-1}K^{-2}. Give an order-of-magnitude estimate of the electronic effective mass in UPt$_3$. Why might UPt$_3$ be known as a 'heavy fermion compound'?

(1.4)* Show that the electrical conductivity σ in the Sommerfeld model at $T = 0$ is given by $\sigma = ne^2\tau/m_e$; i.e. it is identical with that in the Drude model. To do this, first use the relaxation time approximation to show that the effect of an external force f on each electron is to displace the whole Fermi surface by an amount

---
[23] In later chapters we shall see that m^* may be a function of energy and/or a tensor quantity.
[24] There will be more on heterojunctions in later chapters.

$\delta k = f\tau/\hbar$. Hence show that this leads to a total current density in the direction of the force given by

$$J = \frac{2}{(2\pi)^2} \frac{-e\hbar k_F^3}{m_e} \frac{f\tau}{\hbar} \int_0^\pi \sin\theta \cos^2\theta \, d\theta,$$

where θ is the polar angle in spherical polar coordinates. Finally, by putting $f = -eE$, where E is an E-field, obtain the conductivity.

(1.5) Derivation of the thermopower (Thomson coefficient) of a metal within the Sommerfeld model. The thermopower is a voltage which arises because of the temperature difference between two ends of a rod.

a) Show that the chemical potential $\mu(T)$ in the Sommerfeld model has the temperature (T) dependence

$$\mu \approx E_F\{1 - \frac{\pi^2}{12}(\frac{k_B T}{E_F})^2\}, \quad (1.45)$$

where $E_F = \mu(0)$. Evaluate $(E_F - \mu(T))/E_F$ for Cu at room temperature.

b) If one end of the rod is at temperature T_1 and the other is at T_2, then a voltage V will appear across the ends of the rod given by

$$\mu(T_1) - \mu(T_2) = -eV.$$

Hence the thermopower (dV/dT) is given by

$$\frac{dV}{dT} = \frac{1}{e}\frac{d\mu}{dT}.$$

Evaluate this for copper.[25]

(1.6) Derive an expression for the frequency-dependent relative permittivity of a metal using the relaxation-time approximation and the free-electron model. Estimate the plasma frequency for a typical metal such as copper.

(1.7)* (This exercise is a numerical check that the energy broadening ΔE of the Fermi–Dirac distribution function shown in Fig. 1.7 is $\sim k_B T$. It also serves as a test of the approximations in Section 1.5.5.) Evaluate μ for Cu by numerically integrating Equation 1.36, for temperatures 0, 10 K, 30 K, 100 K, 300 K and 1000 K. Use the values obtained to plot the Fermi–Dirac distribution function versus energy for these temperatures (use a spreadsheet or mathematical package), and compare ΔE with $k_B T$ using a criterion of your own devising. (You may ignore the thermal expansion of Cu. On the other hand, it is a rewarding exercise to see how much difference thermal expansion makes to the answer!)

---

[25] Note that the thermopower can only have one sign in the Sommerfeld model, in contrast to observations on real metals, where both positive and negative values are obtained.

2 The quantum mechanics of particles in a periodic potential: Bloch's theorem

2.1 Introduction and health warning

2.1	Introduction and health warning	16
2.2	Introducing the periodic potential	16
2.3	Born–von Karman boundary conditions	17
2.4	The Schrödinger equation in a periodic potential	18
2.5	Bloch's theorem	19
2.6	Electronic bandstructure	20

In this chapter we shall set up the formalism for dealing with a periodic potential; this is known as **Bloch's theorem**. The next three chapters are going to appear to be *hard work* from a conceptual point of view. However, although the algebra *looks* complicated, the underlying ideas are really quite simple; after some reflection, you should be able to reproduce the various derivations yourself.

I am going to justify the Bloch theorem fairly rigorously in this chapter (not *much* cheating). This formalism will then be used to treat two opposite limits, a very weak periodic potential (Chapter 3) and a potential which is so strong that the electrons can hardly move (Chapter 4). You will see that both limits give qualitatively similar answers, i.e. reality, which lies somewhere in between, must also be like this! Once this has been done, I hope that you will believe that there is a good theoretical justification for the real electronic bandstructures which we are going to see and use in subsequent chapters.

This is a good point at which to read Appendix A if you are unfamiliar with (or have forgotten!) the meaning of terms such as *lattice, primitive lattice translation vector* and *unit cell*. We shall be using these terms extensively in the next few pages.

2.2 Introducing the periodic potential

Thus far we have been treating the electrons as totally free. We now wish to introduce the effects of a periodic potential $V(\mathbf{r})$. The underlying translational periodicity of the crystal is defined by the **lattice translation vectors**

$$\mathbf{T} = n_1 \mathbf{a}_1 + n_2 \mathbf{a}_2 + n_3 \mathbf{a}_3, \tag{2.1}$$

where n_1, n_2 and n_3 are integers and $\mathbf{a}_1$, $\mathbf{a}_2$ and $\mathbf{a}_3$ are three noncoplanar vectors (the primitive lattice translation vectors).[1] Now $V(\mathbf{r})$ must be periodic, i.e.

$$V(\mathbf{r} + \mathbf{T}) = V(\mathbf{r}). \tag{2.2}$$

The periodic nature of $V(\mathbf{r})$ also implies that the potential may be expressed as a Fourier series

$$V(\mathbf{r}) = \sum_{\mathbf{G}} V_{\mathbf{G}} e^{i\mathbf{G}\cdot\mathbf{r}}, \tag{2.3}$$

where the $\mathbf{G}$ are a set of vectors and the $V_{\mathbf{G}}$ are Fourier coefficients.[2]

---

[1] See Appendix A or, for example, *Introduction to solid state physics*, by Charles Kittel, seventh edition (Wiley, New York, 1996) pages 4–7.

[2] Notice that the units of $\mathbf{G}$ are the same as those of the wavevector $\mathbf{k}$ (also known as *propagation vector*) of a particle; $|\mathbf{k}| = 2\pi/\lambda$, where λ is the de Broglie wavelength. Therefore the $\mathbf{G}$ are vectors in k-space.

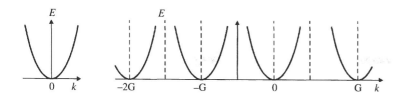

Fig. 2.1 The effect of introducing the reciprocal lattice; instead of dealing with just one electron dispersion relationship (left) we have an infinite number of copies (right). (For those interested in a more mathematically rigorous derivation of the translational symmetry of the wave function in k-space, see Equations 14.24–26 in "Symmetry in Physics", by J. P. Elliott and P. G. Dawber, Oxford University Press, New York, 1979).

Equations 2.2 and 2.3 imply that

$$e^{i\mathbf{G}.\mathbf{T}} = 1, \text{ i.e. } \mathbf{G}.\mathbf{T} = 2l\pi, \quad (2.4)$$

where l is an integer. As $\mathbf{T} = n_1\mathbf{a}_1 + n_2\mathbf{a}_2 + n_3\mathbf{a}_3$, this implies that

$$\mathbf{G} = m_1\mathbf{A}_1 + m_2\mathbf{A}_2 + m_3\mathbf{A}_3, \quad (2.5)$$

where the m_j are integers, and the $\mathbf{A}_j$ are three noncoplanar vectors defined by

$$\mathbf{a}_j.\mathbf{A}_l = 2\pi \delta_{jl}. \quad (2.6)$$

Take a few moments to convince yourself that this is the only way of defining $\mathbf{G}$ which can satisfy eqn 2.4 for all possible $\mathbf{T}$.

Using very simple reasoning, we have shown that the existence of a lattice in r-space automatically implies the existence of a lattice in k-space. The vectors $\mathbf{G}$ define the **reciprocal lattice**; the $\mathbf{A}_j$ are its primitive translation vectors.

The reciprocal lattice has extraordinary consequences for the electronic motion, even before we 'switch on' the lattice potential. Instead of dealing with just one electron dispersion relationship $E(\mathbf{k})$ there must be an infinite number of equivalent dispersion relationships (see Fig. 2.1) such that $E(\mathbf{k}) = E(\mathbf{k}+\mathbf{G})$ for all $\mathbf{G}$ (cf. eqn 2.2). However, the k-space periodicity also implies that all information will be contained in the primitive unit cell of the reciprocal lattice, known as *the first Brillouin zone*.[3] The first Brillouin zone has a k-space volume

$$V_{k3} = \mathbf{A}_1.\mathbf{A}_2 \times \mathbf{A}_3. \quad (2.7)$$

[3]The first Brillouin zone is the **Wigner–Seitz primitive cell** of the reciprocal lattice; see Section A.7 for instructions on constructing such a cell. Just as the primitive unit cell of the r-space lattice contains all of the necessary information about the crystal (it can be translated by the **T**s to fill all space; see Section A6), the first Brillouin zone contains all we need to know about the k-space properties of a solid.

2.3 Born–von Karman boundary conditions

We need to derive a suitable set of functions with which we can describe the motion of the electrons through the periodic potential; 'motion' implies that we do not want standing waves. The functions should reflect the translational symmetry properties of the lattice; to do this we use **Born–von Karman periodic boundary conditions**.

We choose a plane wave

$$\phi(\mathbf{r}) = e^{i(\mathbf{k}.\mathbf{r}-\omega t)} \quad (2.8)$$

subject to boundary conditions which include the symmetry of the crystal

$$\phi(\mathbf{r} + N_j\mathbf{a}_j) = \phi(\mathbf{r}), \quad (2.9)$$

where $j = 1, 2, 3$ and $N = N_1N_2N_3$ is the number of primitive unit cells in the crystal; N_j is the number of unit cells in the jth direction.[4]

[4]It is instructive to compare these with the boundary conditions used in Appendix B (eqn B1).

The boundary condition (eqn 2.9) implies that
$$e^{iN_j\mathbf{k}\cdot\mathbf{a}_j} = 1 \tag{2.10}$$

for $j = 1, 2, 3$. Comparing this with eqn 2.4 (and the discussion that follows it) suggests that the allowed wavevectors are
$$\mathbf{k} = \sum_{j=1}^{3} \frac{m_j}{N_j} \mathbf{A}_j. \tag{2.11}$$

Each time that all of the m_j change by one we generate a new state; therefore the volume of k-space occupied by one state is
$$\frac{\mathbf{A}_1}{N_1} \cdot \frac{\mathbf{A}_2}{N_2} \times \frac{\mathbf{A}_3}{N_3} = \frac{1}{N} \mathbf{A}_1 \cdot \mathbf{A}_2 \times \mathbf{A}_3. \tag{2.12}$$

Comparing this with eqn 2.7 shows that *the Brillouin zone always contains the same number of k-states as the number of primitive unit cells in the crystal.* This fact will be of immense importance later on (remember it!); it will be a key factor in determining whether a material is an insulator, semiconductor or metal.

2.4 The Schrödinger equation in a periodic potential

[5] Note that I have written 'a particle'. The conclusions of this chapter will be seen to be true for *any* particle in *any* periodic potential.

The Schrödinger equation for a particle[5] of mass m in the periodic potential $V(\mathbf{r})$ may be written
$$H\psi = \{-\frac{\hbar^2 \nabla^2}{2m} + V(\mathbf{r})\}\psi = E\psi. \tag{2.13}$$

As before (see eqn 2.3), we write the potential as a Fourier series
$$V(\mathbf{r}) = \sum_{\mathbf{G}} V_{\mathbf{G}} e^{i\mathbf{G}\cdot\mathbf{r}}, \tag{2.14}$$

where the $\mathbf{G}$ are the reciprocal lattice vectors. We are at liberty to set the origin of potential energy wherever we like; as a convenience for later derivations we set the uniform background potential to be zero, i.e.
$$V_0 \equiv 0. \tag{2.15}$$

We can write the wavefunction ψ as a sum of plane waves obeying the Born–von Karman boundary conditions,[6]

[6] That is, the $\mathbf{k}$ in eqn 2.16 are given by eqn 2.11.

$$\psi(\mathbf{r}) = \sum_{\mathbf{k}} C_{\mathbf{k}} e^{i\mathbf{k}\cdot\mathbf{r}}. \tag{2.16}$$

This ensures that ψ also obeys the Born–von Karman boundary conditions.

We now substitute the wavefunction (eqn 2.16) and the potential (eqn 2.14) into the Schrödinger equation (eqn 2.13) to give
$$\sum_{\mathbf{k}} \frac{\hbar^2 k^2}{2m} C_{\mathbf{k}} e^{i\mathbf{k}\cdot\mathbf{r}} + \{\sum_{\mathbf{G}} V_{\mathbf{G}} e^{i\mathbf{G}\cdot\mathbf{r}}\}\{\sum_{\mathbf{k}} C_{\mathbf{k}} e^{i\mathbf{k}\cdot\mathbf{r}}\} = E \sum_{\mathbf{k}} C_{\mathbf{k}} e^{i\mathbf{k}\cdot\mathbf{r}}. \tag{2.17}$$

The potential energy term can be rewritten

$$V(\mathbf{r})\psi = \sum_{\mathbf{G},\mathbf{k}} V_{\mathbf{G}} C_{\mathbf{k}} e^{i(\mathbf{G}+\mathbf{k}).\mathbf{r}}, \qquad (2.18)$$

where the sum on the right-hand side is over all $\mathbf{G}$ and $\mathbf{k}$. As the sum is over all possible values of $\mathbf{G}$ and $\mathbf{k}$, it can be rewritten as[7]

$$V(\mathbf{r})\psi = \sum_{\mathbf{G},\mathbf{k}} V_{\mathbf{G}} C_{\mathbf{k}-\mathbf{G}} e^{i\mathbf{k}.\mathbf{r}}. \qquad (2.19)$$

Therefore the Schrödinger equation (eqn 2.17) becomes

$$\sum_{\mathbf{k}} e^{i\mathbf{k}.\mathbf{r}} \{(\frac{\hbar^2 k^2}{2m} - E) C_{\mathbf{k}} + \sum_{\mathbf{G}} V_{\mathbf{G}} C_{\mathbf{k}-\mathbf{G}} \} = 0. \qquad (2.20)$$

[7] Notice that the $\mathbf{G}$ also obey the Born–von Karman boundary conditions; this may be easily seen if values of m_j that are integer multiples of N_j are substituted in eqn 2.11. As the sum in eqn 2.16 is over all $\mathbf{k}$ that obey the Born–von Karmann boundary conditions, it automatically encompasses all $\mathbf{k} - \mathbf{G}$.

As the Born–von Karman plane waves are an orthogonal set of functions, the coefficient of each term in the sum must vanish (one can prove this by multiplying by a plane wave and integrating), i.e.

$$\left(\frac{\hbar^2 k^2}{2m} - E\right) C_{\mathbf{k}} + \sum_{\mathbf{G}} V_{\mathbf{G}} C_{\mathbf{k}-\mathbf{G}} = 0. \qquad (2.21)$$

(Note that we get the Sommerfeld result if we set $V_{\mathbf{G}} = 0$.)

It is going to be convenient to deal just with solutions in the first Brillouin zone (we have already seen that this contains all useful information about k-space). So, we write $\mathbf{k} = (\mathbf{q} - \mathbf{G}')$, where $\mathbf{q}$ lies in the first Brillouin zone and $\mathbf{G}'$ is a reciprocal lattice vector. Equation 2.21 can then be rewritten

$$\left(\frac{\hbar^2 (\mathbf{q}-\mathbf{G}')^2}{2m} - E\right) C_{\mathbf{q}-\mathbf{G}'} + \sum_{\mathbf{G}} V_{\mathbf{G}} C_{\mathbf{q}-\mathbf{G}'-\mathbf{G}} = 0. \qquad (2.22)$$

Finally, we change variables so that $\mathbf{G}'' \to \mathbf{G} + \mathbf{G}'$, leaving the equation of coefficients in the form

$$\left(\frac{\hbar^2 (\mathbf{q}-\mathbf{G}')^2}{2m} - E\right) C_{\mathbf{q}-\mathbf{G}'} + \sum_{\mathbf{G}''} V_{\mathbf{G}''-\mathbf{G}'} C_{\mathbf{q}-\mathbf{G}''} = 0. \qquad (2.23)$$

This equation of coefficients is very important, in that it specifies the $C_{\mathbf{k}}$ which are used to make up the wavefunction ψ in eqn 2.16.

Thus far, I have merely been performing algebraic manipulations (changing the index of summations and so on); no 'magic' has been involved, and I have tried to keep the notation as simple as I can. However, it is hard to absorb algebra by just looking at it. If you want to understand this material, I recommend that you work through the steps from eqn 2.11 to eqn 2.23 on your own on a piece of paper.

2.5 Bloch's theorem

Equation 2.23 only involves coefficients $C_{\mathbf{k}}$ in which $\mathbf{k} = \mathbf{q} - \mathbf{G}$, with the $\mathbf{G}$ being general reciprocal lattice vectors. In other words, if we choose a particular value of **q**, then the only $C_\mathbf{k}$ that feature in eqn 2.23 are of the form $C_{\mathbf{q}-\mathbf{G}}$; these coefficients specify the form that the wavefunction ψ will take (see eqn 2.16).

Therefore, for each distinct value of **q**, there is a wavefunction $\psi_\mathbf{q}(\mathbf{r})$ that takes the form

$$\psi_\mathbf{q}(\mathbf{r}) = \sum_\mathbf{G} C_{\mathbf{q}-\mathbf{G}} e^{i(\mathbf{q}-\mathbf{G})\cdot\mathbf{r}}, \tag{2.24}$$

where we have obtained the equation by substituting $\mathbf{k} = \mathbf{q} - \mathbf{G}$ into eqn 2.16. Equation 2.24 can be rewritten

$$\psi_\mathbf{q}(\mathbf{r}) = e^{i\mathbf{q}\cdot\mathbf{r}} \sum_\mathbf{G} C_{\mathbf{q}-\mathbf{G}} e^{-i\mathbf{G}\cdot\mathbf{r}} = e^{i\mathbf{q}\cdot\mathbf{r}} u_{j,\mathbf{q}}, \tag{2.25}$$

i.e. (a plane wave with wavevector within the first Brillouin zone)×(a function $u_{j,\mathbf{q}}$ with the periodicity of the lattice).[8]

This leads us to **Bloch's theorem**. 'The eigenstates ψ of a one-electron Hamiltonian $H = -\frac{\hbar^2 \nabla^2}{2m} + V(\mathbf{r})$, where $V(\mathbf{r}+\mathbf{T}) = V(\mathbf{r})$ for all Bravais lattice[9] translation vectors **T** can be chosen to be a plane wave times a function with the periodicity of the Bravais lattice.' An alternative (algebraic) statement of Bloch's law will be given in eqn 4.3 below.

Note that Bloch's theorem

- is true for *any* particle propagating in a lattice (even though Bloch's theorem is traditionally stated in terms of electron states (as above), in the derivation we made no assumptions about what the particle *was*);
- makes no assumptions about the *strength* of the potential.

[8] You can check that $u_{j,\mathbf{q}}$ (i.e. the sum in eqn 2.25) has the periodicity of the lattice by making the substitution $\mathbf{r} \to \mathbf{r} + \mathbf{T}$ (see eqns 2.2, 2.3 and 2.4 if stuck).

[9] See Section A.11 for a definition of the Bravais lattices.

2.6 Electronic bandstructure

Let us focus specifically on electrons in a periodic potential. Equation 2.25 hints at the idea of **electronic bandstructure**. Each set of $u_{j,\mathbf{q}}$ is going to result in a set of electron states with a particular character (e.g. whose energies lie on a particular dispersion relationship[10]); this is the basis of our idea of an **electronic band**. The number of possible wavefunctions in this band is just going to be given by the number of distinct **q**, i.e. the number of Born–von Karman wavevectors in the first Brillouin zone. Therefore the number of electron states in each band is just 2×(the number of primitive cells in the crystal), where the factor two has come from spin-degeneracy (see eqn 2.12). This is going to be very important in our ideas about **band filling**, and the classification of materials into metals, semimetals, semiconductors and insulators.

We are now going to consider two tractable limits of Bloch's theorem, a very weak periodic potential and a very strong periodic potential (so strong that the electrons can hardly move from atom to atom). We shall see that both extreme limits give rise to *bands*, with *band gaps* between them. In both extreme cases, the bands are qualitatively very similar; i.e. real potentials, which must lie somewhere between the two extremes, must also give rise to qualitatively similar bands and band gaps.

[10] A dispersion relationship describes the relationship between a particle's energy E and its **q**; it is the function $E = E(\mathbf{q})$.

Further reading

A simple justification of the Bloch theorem is given in *Introduction to solid state physics*, by Charles Kittel, seventh edition (Wiley, New York, 1996) in the first few pages of Chapter 7 and in *Solid state physics*, by G. Burns (Academic Press, Boston, 1995) Section 10.4; an even more elementary one is found in *Electricity and magnetism*, by B.I. Bleaney and B. Bleaney, third edition (Oxford University Press, Oxford, 1989) Section 12.3. More rigorous and general proofs are available in *Solid state physics*, by N.W Ashcroft and N.D. Mermin (Holt, Rinehart and Winston, New York, 1976) pages 133–140.

An alternative (and very general) proof of Bloch's theorem is given in Chapter 1 of *Quantum theory of the solid state*, by Joseph Callaway (Academic Press, San Diego, 1991).

Exercises

(2.1) Justify the following form for the reciprocal lattice vectors in a three-dimensional crystal

$$\mathbf{A}_1 = \frac{2\pi(\mathbf{a}_2 \times \mathbf{a}_3)}{\mathbf{a}_1 \cdot (\mathbf{a}_2 \times \mathbf{a}_3)}.$$

($\mathbf{A}_2$ and $\mathbf{A}_3$ may be obtained by permuting the indices.)

Hence, use eqn 2.7 to evaluate the Brillouin-zone volume[11] in terms of $\mathbf{a}_1$, $\mathbf{a}_2$ and $\mathbf{a}_3$.

Finally, show, using eqn 1.19, that the first Brillouin zone of a crystal containing N real-space unit cells will contain exactly N **k**-states.

(2.2) Show that the spacing $d(m_1, m_2, m_3)$ of lattice planes with the Miller indices (m_1, m_2, m_3) (see Appendix A) is given by

$$d(m_1, m_2, m_3) = \frac{2\pi}{|\mathbf{G}(m_1, m_2, m_3)|},$$

where $\mathbf{G}(m_1, m_2, m_3) = m_1\mathbf{A}_1 + m_2\mathbf{A}_2 + m_3\mathbf{A}_3$.

(2.3)* This question is intended to help you understand the way in which real bands are plotted in later chapters. It shows that in even quite a simple situation, a diagram which is at first sight rather daunting can result! Figure 2.2 shows a plan view of the Brillouin zone of a square lattice, with some points of high symmetry labelled. Sketch the energies of the free-electron bands $E = (\hbar^2(\mathbf{k} - \mathbf{G})^2)/(2m_e)$ up to $E \sim 10\hbar^2/8m_ea^2$ as k traverses the path $\Gamma - M - X - \Gamma$ along the straight lines shown. You will need to consider dispersion curves originating from several neighbouring Brillouin zones.

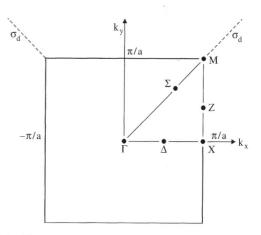

Fig. 2.2 A plan view of the Brillouin zone of a square lattice, with some points of high symmetry labelled.

It will become evident that dispersion curves converge at points of high symmetry; label the **degeneracy**, i.e. the number of curves that converge, at such places on your diagram.

(2.4) Look up the $\mathbf{a}_j$ for the body-centred cubic (bcc) and face-centred cubic (fcc) lattices (see Figs. A.2 and A.3). Show that the reciprocal lattice of the bcc lattice is fcc and vice versa. For both bcc and fcc work out the magnitude of $|\mathbf{A}_j|$ in terms of a, the edge of the conventional cubic unit cell.

---

[11] You may need the vector identity $\mathbf{P} \times \mathbf{Q} \times \mathbf{R} = \mathbf{Q}(\mathbf{P}\cdot\mathbf{R}) - \mathbf{R}(\mathbf{P}\cdot\mathbf{Q})$. For a quick revision of vectors, see, for example, *Electricity and magnetism*, by B.I. Bleaney and B. Bleaney, third edition (Oxford University Press, Oxford, 1989), Appendix A.

(2.5) The two-dimensional square lattice has $\mathbf{a}_1 = a\mathbf{e}_1$ and $\mathbf{a}_2 = a\mathbf{e}_2$. Show that the reciprocal lattice is also square. Use the following recipe for finding the first three Brillouin zones (see Section A.7). *The Brillouin zone is obtained by starting at any reciprocal lattice point (which is then the origin) and constructing vectors to all neighbouring reciprocal lattice points. Lines are then constructed perpendicular to and passing through the midpoints of these vectors. The Brillouin zone is the cell with the smallest area about the origin bounded by these lines. The second Brillouin zone is the area between the first Brillouin zone and the next smallest closed polygon formed by the lines. Similarly, the third Brillouin zone is the area between the second Brillouin zone and the next closed polygon formed by the lines.*

Show that the first Brillouin zone is a square, the second is a set of four isoceles triangles, and the third is a set of eight isoceles triangles. Show that the first three Brillouin zones have the same area. (Hint: see Fig. 3.4(a) if stuck.)

The nearly-free electron model

3.1 Introduction

We are now going to consider the solutions of the Schrödinger equation for an electron in a periodic potential in the case that the potential is very weak, i.e. the electron is *nearly free*. For algebraic convenience, let us write the free-electron energy of a state with $\mathbf{k} = \mathbf{q} - \mathbf{G}$ as $E^0_{\mathbf{q}-\mathbf{G}} = \frac{\hbar^2(\mathbf{q}-\mathbf{G})^2}{2m}$ (see Fig. 2.1). We substitute this in eqn 2.23 to give

3.1	Introduction	23
3.2	Vanishing potential	23
3.3	Consequences of the nearly-free-electron model	26

$$(E^0_{\mathbf{q}-\mathbf{G}'} - E)C_{\mathbf{q}-\mathbf{G}'} + \sum_{\mathbf{G}''} V_{\mathbf{G}''-\mathbf{G}'}C_{\mathbf{q}-\mathbf{G}''} = 0. \qquad (3.1)$$

We separate the potential term into two parts

$$(E^0_{\mathbf{q}-\mathbf{G}'} - E)C_{\mathbf{q}-\mathbf{G}'} + V_{\mathbf{G}'-\mathbf{G}'}C_{\mathbf{q}-\mathbf{G}'} + \sum_{\mathbf{G}'' \neq \mathbf{G}'} V_{\mathbf{G}''-\mathbf{G}'}C_{\mathbf{q}-\mathbf{G}''} = 0. \qquad (3.2)$$

Now $V_{\mathbf{G}'-\mathbf{G}'} \equiv V_0$; we defined $V_0 \equiv 0$ (see eqn 2.15), so that this equation becomes

$$(E^0_{\mathbf{q}-\mathbf{G}'} - E)C_{\mathbf{q}-\mathbf{G}'} + \sum_{\mathbf{G}'' \neq \mathbf{G}'} V_{\mathbf{G}''-\mathbf{G}'}C_{\mathbf{q}-\mathbf{G}''} = 0. \qquad (3.3)$$

Thus far we have only been fiddling about algebraically; with the equation of coefficients in this form, we can now start to examine the behaviour of the solutions as the potential becomes very small.

3.2 Vanishing potential

3.2.1 Single electron energy state

Consider an electron energy state defined by $E(\mathbf{q} - \mathbf{G}')$, which is such that there are no other states at the same energy at that point in k-space. We then take eqn 3.3 and let $V_\mathbf{G} \to 0$, giving

$$(E^0_{\mathbf{q}-\mathbf{G}'} - E)C_{\mathbf{q}-\mathbf{G}'} = 0, \qquad (3.4)$$

i.e. either

$$E^0_{\mathbf{q}-\mathbf{G}'} - E = 0 \qquad (3.5)$$

or

$$C_{\mathbf{q}-\mathbf{G}'} = 0. \qquad (3.6)$$

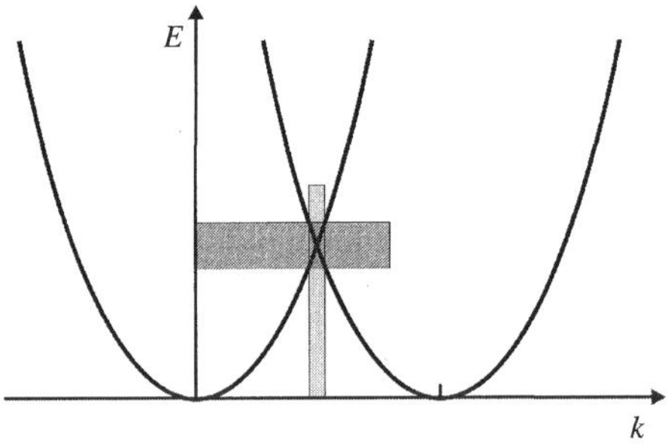

Fig. 3.1 Repeated electron dispersion relationships crossing at the Brillouin-zone boundary. Shading indicates schematically the regions where the energies of the curves are within $|V|$ of each other.

However, eqn 3.5 defines the electron energy state under consideration, so that $C_{\mathbf{q}-\mathbf{G}'}$ must be finite for this state and zero for all other cases, i.e.

$$E(\mathbf{q}-\mathbf{G}') = E^0_{\mathbf{q}-\mathbf{G}'} = \frac{\hbar^2(\mathbf{q}-\mathbf{G}')^2}{2m}, \quad (3.7)$$

with the spatial part of the wavefunction being given by

$$\psi \propto e^{i(\mathbf{q}-\mathbf{G}')\cdot\mathbf{r}}. \quad (3.8)$$

In other words, the solution is a single plane wave.

The above manipulation has, of course, given us back the free-electron result (with the added ingredient of the periodicity of k-space). Whilst this is reassuring and perhaps even trivially obvious, it will serve as a very useful contrast to what happens in the following section. There is, however, another, much more important implication; note that $C_{\mathbf{q}-\mathbf{G}} = 0$ for $\mathbf{G} \neq \mathbf{G}'$. The vanishing of all the $C_{\mathbf{q}-\mathbf{G}}$ which have $\mathbf{G} \neq \mathbf{G}'$ as $V_{\mathbf{G}} \to 0$, implies that the sum term of eqn 3.1 only causes corrections to the energy which are second order in the potential V (and therefore minute) when V is finite but small.

Therefore, in the case of well-separated free-electron states and a weak potential, eqn 3.3 can be rewritten

$$(E^0_{\mathbf{q}-\mathbf{G}'} - E)C_{\mathbf{q}-\mathbf{G}'} + O(V^2) = 0, \quad (3.9)$$

i.e. the corrections to the free-electron energies are going to be small.

3.2.2 Several degenerate energy levels

Let us now assume that several different free-electron dispersion relationships cross at a particular point in k space (by symmetry such crossings will tend to occur at Brillouin-zone boundaries; see Fig. 3.1). Equation 3.5 can then be true simultaneously for several different levels. This means that a number (i.e. not just one) of $C_{\mathbf{q}-\mathbf{G}'}$ can be arbitrarily large and the wavefunction will become a linear combination of plane waves. We now take a special case of this, turn on the potential and see what happens.

3.2.3 Two degenerate free-electron levels

We take two dispersion curves $E^0_{\mathbf{q}-\mathbf{G}_1}$ and $E^0_{\mathbf{q}-\mathbf{G}_2}$ at a point in k-space where their energy separation is less than $|V|$, i.e.

$$|E^0_{\mathbf{q}-\mathbf{G}_1} - E^0_{\mathbf{q}-\mathbf{G}_2}| \leq V \quad (3.10)$$

(see Fig. 3.1). For all other $\mathbf{G} \neq \mathbf{G_2}$, however,

$$|E^0_{\mathbf{q}-\mathbf{G_1}} - E^0_{\mathbf{q}-\mathbf{G}}| \gg V. \qquad (3.11)$$

We rewrite eqn 3.1 for the two levels as follows

$$(E - E^0_{\mathbf{q}-\mathbf{G_1}})C_{\mathbf{q}-\mathbf{G_1}} =$$
$$V_{\mathbf{G_1}-\mathbf{G_1}}C_{\mathbf{q}-\mathbf{G_1}} + V_{\mathbf{G_2}-\mathbf{G_1}}C_{\mathbf{q}-\mathbf{G_2}} + \sum_{\mathbf{G} \neq \mathbf{G_1},\mathbf{G_2}} V_{\mathbf{G}-\mathbf{G_1}}C_{\mathbf{q}-\mathbf{G}} \qquad (3.12)$$

and

$$(E - E^0_{\mathbf{q}-\mathbf{G_2}})C_{\mathbf{q}-\mathbf{G_2}} =$$
$$V_{\mathbf{G_2}-\mathbf{G_2}}C_{\mathbf{q}-\mathbf{G_2}} + V_{\mathbf{G_1}-\mathbf{G_2}}C_{\mathbf{q}-\mathbf{G_1}} + \sum_{\mathbf{G} \neq \mathbf{G_1},\mathbf{G_2}} V_{\mathbf{G}-\mathbf{G_2}}C_{\mathbf{q}-\mathbf{G}}. \qquad (3.13)$$

As before, $V_{\mathbf{G_1}-\mathbf{G_1}} \equiv V_{\mathbf{G_2}-\mathbf{G_2}} \equiv V_0 \equiv 0$ (see eqn 2.15). Furthermore, using similar arguments to those in Section 3.2.1, the $\sum$ terms in eqns 3.12 and 3.13 will be second order in V and therefore very small in a weak potential. Ignoring the small $\sum$ terms, eqns 3.12 and 3.13 become

$$(E - E^0_{\mathbf{q}-\mathbf{G_1}})C_{\mathbf{q}-\mathbf{G_1}} = V_{\mathbf{G_2}-\mathbf{G_1}}C_{\mathbf{q}-\mathbf{G_2}} \qquad (3.14)$$

and

$$(E - E^0_{\mathbf{q}-\mathbf{G_2}})C_{\mathbf{q}-\mathbf{G_2}} = V_{\mathbf{G_1}-\mathbf{G_2}}C_{\mathbf{q}-\mathbf{G_1}}. \qquad (3.15)$$

Notice that *two* coefficients, $C_{\mathbf{q}-\mathbf{G_1}}$ and $C_{\mathbf{q}-\mathbf{G_2}}$, are finite, so that the solution is now a **superposition of two plane waves** (cf. the single plane wave of the well-separated energy levels of the previous section).[1] We now simplify the notation by putting $\mathbf{k} = \mathbf{q} - \mathbf{G_1}$ and $\mathbf{G} = \mathbf{G_2} - \mathbf{G_1}$ (as the difference of two reciprocal lattice vectors is also a reciprocal lattice vector) to give

$$(E - E^0_{\mathbf{k}})C_{\mathbf{k}} = V_{\mathbf{G}}C_{\mathbf{k}-\mathbf{G}} \qquad (3.16)$$

and

$$(E - E^0_{\mathbf{k}-\mathbf{G}})C_{\mathbf{k}-\mathbf{G}} = V_{-\mathbf{G}}C_{\mathbf{k}} = V^*_{\mathbf{G}}C_{\mathbf{k}}, \qquad (3.17)$$

where we have used the fact that the potential must be real to give $V_{-\mathbf{G}} = V^*_{\mathbf{G}}$. Equations 3.16 and 3.17 are simultaneous equations of coefficients $C_{\mathbf{k}}$ and $C_{\mathbf{k}-\mathbf{G}}$; the solution is

$$E = \frac{1}{2}(E^0_{\mathbf{k}} + E^0_{\mathbf{k}-\mathbf{G}}) \pm \{(\frac{E^0_{\mathbf{k}} - E^0_{\mathbf{k}-\mathbf{G}}}{2})^2 + |V_{\mathbf{G}}|^2\}^{\frac{1}{2}}, \qquad (3.18)$$

i.e.

$$E \approx \frac{1}{2}(E^0_{\mathbf{k}} + E^0_{\mathbf{k}-\mathbf{G}}) \pm |V_{\mathbf{G}}|. \qquad (3.19)$$

Therefore we have a symmetrical splitting of levels about the free electron value close to where $E^0_{\mathbf{k}}$ and $E^0_{\mathbf{k}-\mathbf{G}}$ cross (see Fig. 3.2). Now $E^0_{\mathbf{k}} = E^0_{\mathbf{k}-\mathbf{G}}$ when $|\mathbf{k}| = |\mathbf{k} - \mathbf{G}|$, i.e. $\mathbf{k}$ lies on a plane half way between 0 and $\mathbf{G}$. The vector $\mathbf{k} - \frac{1}{2}\mathbf{G}$ will be parallel to this plane, so that $\mathbf{G}.(\mathbf{k} - \frac{1}{2}\mathbf{G}) = 0$, i.e.

$$G^2 = 2\mathbf{k}.\mathbf{G}. \qquad (3.20)$$

We can say that a weak periodic potential has its strongest effect on states with

[1] It is illuminating to look at the two waves involved, $\psi_1 = C_{\mathbf{q}-\mathbf{G_1}}e^{i(\mathbf{q}-\mathbf{G_1}).\mathbf{r}}$ and $\psi_2 = C_{\mathbf{q}-\mathbf{G_2}}e^{i(\mathbf{q}-\mathbf{G_2}).\mathbf{r}}$. The probability density functions of the two superimposed waves are $|\Psi_\pm|^2 = |\psi_1 \pm \psi_2|^2$. If we assume (for simplicity) that $C_{\mathbf{q}-\mathbf{G_1}} = C_{\mathbf{q}-\mathbf{G_2}} = C$, then $|\Psi_+|^2 = |\psi_1 + \psi_2|^2 = 2|C|^2(1+\cos((\mathbf{G_1}-\mathbf{G_2}).\mathbf{r})$. Now, as $\mathbf{G_1}$ and $\mathbf{G_2}$ are reciprocal lattice vectors, then $(\mathbf{G_1}-\mathbf{G_2})$ is also a reciprocal lattice vector. $|\Psi_+|^2$ therefore contains a standing-wave component which exhibits one of the real-space periodicities of the lattice; hence the electron wavefunction will tend to 'lock on' to this periodicity. Similarly, $|\Psi_-|^2 = |\psi_1 - \psi_2|^2 = 2|C|^2(1-\cos((\mathbf{G_1}-\mathbf{G_2}).\mathbf{r})$ produces a standing wave component of the same periodicity *but π radians out of phase*. Hence the electron wavefunction can lock onto the same real-space periodicity with either phase (that shown by Ψ_+ or that shown by Ψ_-), leading to the two different possible energies in eqn 3.19; Ψ_+ will tend to concentrate the electron probability density close to the ionic cores (lower energy). Ψ_- will tend to concentrate it between the ionic cores (higher energy).

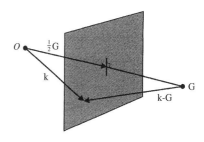

Fig. 3.2 The plane defining the points at which $E^0_{\mathbf{k}} = E^0_{\mathbf{k}-\mathbf{G}}$.

26 *The nearly-free electron model*

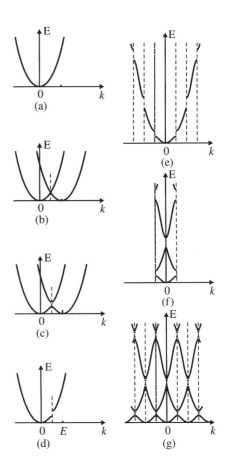

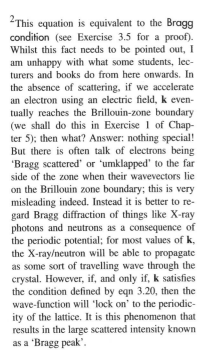

Fig. 3.3 The evolution of nearly-free-electron bands: (a) initial free-electron dispersion relationship; (b) replication of free-electron dispersion curves due to periodicity of k-space (see Chapter 2); (c) opening up at a band gap at the point on the Brillouin-zone boundary where two free-electron curves cross; (d) the effect on the free-electron dispersion relationship of a band gap opening at each successive Brillouin zone boundary. The periodicity of k-space means that the curves shown in (d) must be present in *every* Brillouin zone; (e), (f) and (g) show different ways of representing the resulting bandstructure: (e) is the **extended-zone** scheme, in which each Brillouin zone has one band shown, (f) is the **reduced-zone** scheme whereby all bands are shown within the first Brillouin zone; (g) is the *repeated-zone* or **periodic-zone** scheme.

---

[2] This equation is equivalent to the **Bragg condition** (see Exercise 3.5 for a proof). Whilst this fact needs to be pointed out, I am unhappy with what some students, lecturers and books do from here onwards. In the absence of scattering, if we accelerate an electron using an electric field, **k** eventually reaches the Brillouin-zone boundary (we shall do this in Exercise 1 of Chapter 5); then what? Answer: nothing special! But there is often talk of electrons being 'Bragg scattered' or 'umklapped' to the far side of the zone when their wavevectors lie on the Brillouin zone boundary; this is very misleading indeed. Instead it is better to regard Bragg diffraction of things like X-ray photons and neutrons as a consequence of the periodic potential; for most values of **k**, the X-ray/neutron will be able to propagate as some sort of travelling wave through the crystal. However, if, and only if, **k** satisfies the condition defined by eqn 3.20, then the wave-function will 'lock on' to the periodicity of the lattice. It is this phenomenon that results in the large scattered intensity known as a 'Bragg peak'.

wavevectors close to the condition defined by eqn 3.20.[2] At these points, a **band-gap** opens up symmetrically about the free-electron dispersion relationships. The resulting energy bands are shown in a variety of representations in Fig. 3.3(a)–(f), which also summarises the steps we have followed in the derivation above and introduces the **repeated-zone**, **extended-zone** and **reduced-zone** representations of bandstructure.

3.3 Consequences of the nearly-free-electron model

We have derived two simple rules, which are

- away from Brillouin-zone boundaries the electronic bands (i.e. dispersion relationships) are very similar to those of a free electron;
- bandgaps open up whenever $E(\mathbf{k})$ surfaces cross, which means in particular at the zone boundaries.

To see how these rules influence the properties of real metals, we must remember that each band in the Brillouin zone will contain $2N$ electron states (see Sections 2.3 and 2.6), where N is the number of primitive unit cells in the crystal. We now discuss a few specific cases.[3]

3.3.1 The alkali metals

The alkali metals Na, K *et al.* are monovalent (i.e. have one electron per primitive cell). As a result, their Fermi surfaces, encompassing N states, have a volume which is half that of the first Brillouin zone. Let us examine the geometry of this situation a little more closely.

The alkali metals have a body-centred cubic lattice with a basis comprising a single atom. The conventional unit cell of the body-centred cubic lattice is a cube of side a containing two lattice points (and hence 2 alkali metal atoms) (see Section A.11). The electron density is therefore $n = 2/a^3$. Substituting this in the equation for free-electron Fermi wavevector (eqn 1.23) we find $k_\mathrm{F} = 1.24\pi/a$. The shortest distance to the Brillouin zone boundary is half the length of one of the $\mathbf{A}_j$ for the body-centred cubic lattice (see Exercise 2.4), which is

$$\frac{1}{2}\frac{2\pi}{a}(1^2 + 1^2 + 0^2)^{\frac{1}{2}} = 1.41\frac{\pi}{a}.$$

Hence the free-electron Fermi-surface reaches only $1.24/1.41 = 0.88$ of the way to the closest Brillouin-zone boundary. The populated electron states therefore have ks which lie well clear of any of the Brillouin-zone boundaries, thus avoiding the distortions of the band due to the bandgaps; hence, the alkali metals have properties which are quite close to the predictions of the Sommerfeld model (e.g. a Fermi surface which is spherical to one part in 10^3).[4,5]

3.3.2 Elements with even numbers of valence electrons

These substances contain just the right number of electrons ($2Nj$) to completely fill an integer number j bands up to a band gap. The gap will energetically separate completely filled states from the next empty states; to drive a net current through such a system, one must be able to change the velocity of an electron, i.e move an electron into an unoccupied state of different velocity. However, there are no easily accessible empty states so that such substances should not conduct electricity at $T = 0$; at finite temperatures, electrons will be thermally excited across the gap, leaving filled and empty states in close energetic proximity both above and below the gap so that electrical conduction can occur. Diamond (an insulator), Ge and Si (semiconductors) are good examples.

However, the divalent metals Ca *et al.* plainly conduct electricity rather well. To see why this is the case, consider the Fermi surface of the two-dimensional divalent metal with a square lattice shown in Fig. 3.4. Initially, the free-electron Fermi surface is a circle with an area equivalent to the first Brillouin zone (Fig. 3.4(a)), which consequently straddles the Brillouin-zone boundary (Figs. 3.4(b) and (c)). Figs. 3.4 (d) and (e) show what happens when a weak periodic potential is 'turned on' and band-gaps open up at the Brillouin-zone boundaries; the band-gap raises the energy of the states close to the zone

[3] The aim here is to promote a *general* understanding. If more details of specific cases are required, see e.g. *Solid state physics*, by N.W Ashcroft and N.D. Mermin (Holt, Rinehart and Winston, New York, 1976) Chapter 15.

[4] See D. Shoenberg, *Proc. int. conf. low temperature physics LT9* (Plenum, New York, 1965).

[5] Many books discuss monovalent metals using a one-dimensional model. This is slightly unfortunate, as we shall see in Section 7.6 that monovalent one-dimensional metals attempt to become insulators at low temperatures!

Fig. 3.4 The evolution of the Fermi surface of a divalent two-dimensional metal with a square lattice as a band gap is opened at the Brillouin zone boundary: (a) free-electron Fermi surface (shaded circle), reciprocal lattice points (solid dots) and first (square) second (four isoceles triangles) and third (eight isoceles triangles) Brillouin zones (see Exercise 2.5); (b) the section of Fermi surface enclosed by the first Brillouin zone; (c) the sections of Fermi surface in the second Brillouin zone; (d) distortion of the Fermi-surface section shown in (a) due to formation of band gaps at the Brillouin-zone boundaries; (e) the section of the distorted Fermi surface enclosed by the first Brillouin zone; (f) result of the distortion of the Fermi-surface section in (c) plus 'folding back' of these sections due to the periodicity of k-space.

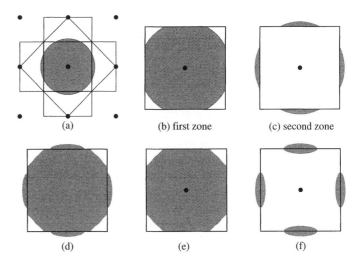

edge in Fig. 3.4(c) and lowers those close to the zone edge in Fig. 3.4(b) (see Fig. 3.3). For ease of reference, we shall call the former states 'the upper band' and the latter states 'the lower band'. Hence some electrons will transfer back from the upper band (the states above the gap) to the lower band (the states below it), tending to distort the Fermi surface sections close to the Brillouin-zone boundaries.

In the situation shown in Figs. 3.4(d), (e), and (f), the material is obviously still an electrical conductor, as filled and empty states are adjacent in energy. Let us call the band-gap at the centres of the Brillouin-zone edges $E_{\text{g}}^{\text{cent}}$ and that at the corners of the Brillouin zone $E_{\text{g}}^{\text{corn}}$.[6] The lowest energy states in the upper band will be at points $(\pm\frac{\pi}{a}, 0)$, $(0, \pm\frac{\pi}{a})$, where a is the lattice parameter of the square lattice, with energy

$$E_{\text{lowest}}^{\text{u}} = \frac{\hbar^2}{2m_{\text{e}}}\frac{\pi^2}{a^2} + \frac{E_{\text{g}}^{\text{cent}}}{2}, \quad (3.21)$$

i.e. the free-electron energy plus half the energy gap. Similarly, the highest energy states in the lower band will be at the points $(\pm\frac{\pi}{a}, \pm\frac{\pi}{a})$, with energy

$$E_{\text{highest}}^{\text{l}} = \frac{\hbar^2}{2m_{\text{e}}}\frac{2\pi^2}{a^2} - \frac{E_{\text{g}}^{\text{corn}}}{2}, \quad (3.22)$$

i.e. the free-electron energy minus half the energy gap. Therefore the material will be a conductor as long as $E_{\text{highest}}^{\text{l}} > E_{\text{lowest}}^{\text{u}}$. Only if the band gap is big enough for $E_{\text{highest}}^{\text{l}} < E_{\text{lowest}}^{\text{u}}$, will all of the electrons be in the lower band at $T = 0$, which will then be completely filled; filled and empty states will be separated in energy by a gap and the material will be an insulator at $T = 0$.

Thus, in general, in two- and three-dimensional divalent metals, the geometrical properties of the dispersion relationships allow the highest states of the lower band (at the corners of the first Brillouin zone boundary most distant from the zone centre) to be at a higher energy than the lowest states of the upper band (at the closest points on the zone boundary to the zone centre). Therefore both bands are partly filled, ensuring that such substances conduct electricity

---

[6] $E_{\text{g}}^{\text{cent}}$ and $E_{\text{g}}^{\text{corn}}$ will in general not be the same; two plane waves contribute to the former and four to the latter. However, the gaps will be of similar magnitude.

at $T = 0$ (see also Exercise 3.4); only one-dimensional divalent metals have no option but to be insulators.

We shall see in Chapter 5 that the empty states at the top of the lowest band (e.g. the unshaded states in the corners in Fig. 3.4(d)) act as *holes* behaving as though they have a *positive* charge. This is the reason for the positive Hall coefficients observed in many divalent metals (see Table 1.1).

3.3.3 More complex Fermi surface shapes

The Fermi surfaces of many simple di- and trivalent metals can be understood adequately by the following sequence of processes.

(1) Construct a free-electron Fermi sphere corresponding to the number of valence electrons.
(2) Construct a sufficient number of Brillouin zones to enclose the Fermi sphere.
(3) Split and 'round off' the edges of the Fermi surface wherever it cuts a Brillouin-zone boundary (i.e. points at which band gaps opens up).
(4) Apply the periodicity of k-space by replicating all of the Fermi-surface sections at equivalent points in the first Brillouin zone.

These steps are illustrated in great detail for a number of cases in e.g. *Solid state physics*, by N.W Ashcroft and N.D. Mermin (Holt, Rinehart and Winston, New York, 1976), Chapter 9 or *Quantum theory of solids*, by C. Kittel, (Wiley, New York, 1987), Chapter 13. I shall merely remark here that the shape of the Brillouin zone has a profound effect on the Fermi surface sections generated by this process. This is simply illustrated by the Fermi surface shown in Fig. 3.5. As in Fig. 3.4, we have a divalent metal. However, in Fig. 3.5, the Brillouin zone is *rectangular*, rather than square, so that the Fermi surface only cuts two

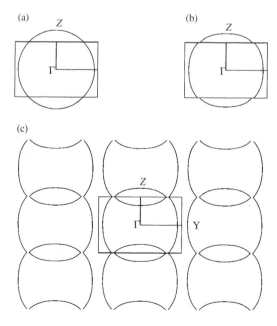

Fig. 3.5 The evolution of the Fermi surface of a divalent two-dimensional metal as a band-gap is opened at the Brillouin zone boundary. (a) Free-electron Fermi circle and rectangular Brillouin zone; (b) the effect of a band-gap opening up at the Brillouin-zone boundary; (c) resulting Fermi surface sections in the extended-zone scheme.

of the Brillouin zone edges. Thus, after band gaps have opened up, the Fermi surface consists of a closed ellipse plus corrugated open lines, rather than the two closed sections of Fig. 3.4.

Further reading

A more detailed treatement of traditional elemental metals is given in *Solid state physics*, by N.W Ashcroft and N.D. Mermin (Holt, Rinehart and Winston, New York, 1976), Chapters 8, 9, 10 and 12 (even if you understand nothing of the discussion, the pictures are good). Simpler discussions are available in *Electricity and magnetism*, by B.I. Bleaney and B. Bleaney, third edition (Oxford University Press, Oxford, 1989), Chapter 12, *Solid state physics*, by G. Burns (Academic Press, Boston, 1995), Sections 10.1–10.21, *Electrons in metals and semiconductors*, by R.G. Chambers (Chapman and Hall, London, 1990) Chapters 4–6, *Introduction to solid state physics*, by Charles Kittel, seventh edition (Wiley, New York, 1996), Chapters 8 and 9.

Pictures of the Fermi surfaces of elemental metals, along with relevant references, are provided in *Electronic theory of metals*, by I.M. Lifshitz, M.Ya. Azbel and M.I. Kaganov (Consultants Bureau, New York, 1973). See also P.B. Visscher and L.M. Falicov, *Phys. Stat. Sol.* B **54**, 9 (1972).

Exercises

(3.1) Work through the algebraic steps required to produce eqn 3.19. Start with eqns 3.16 and 3.17.

(3.2) a) A two-dimensional solid has a rectangular lattice with primitive lattice translation vectors $a\mathbf{e_1}$ and $b\mathbf{e_2}$ ($\mathbf{e_1}$ and $\mathbf{e_2}$ are the Cartesian unit vectors along x and y), with $b > a$. There is one divalent atom per unit cell. Electrons in the solid experience a weak periodic potential $V(x, y)$, with

$$V(x, y) = V_0 + v[\cos(\frac{2\pi x}{a}) + \cos(\frac{2\pi y}{b})]. \quad (3.23)$$

Draw the first and second Brillouin zones. What sort of eigenfunctions are expected for electrons with wavevector $\mathbf{k}$ near the zone centre? Why are electronic wavefunctions of the form $\cos(\frac{\pi x}{a})$ and $\sin(\frac{\pi x}{a})$ appropriate for electrons with $\mathbf{k} = (\frac{\pi}{a}, 0)$? Calculate the energy gap at this point on the first Brillouin zone boundary from first principles.

b) Show how an increase in the parameter v can change the solid from a metal into an insulator and estimate the value of $v = v_{\text{MI}}$ at which this occurs. (Hint: compare the energy of the lower band at the corner of the Brillouin zone with the energy of the lowest point in the upper band.)

c) Sketch how the Fermi surface evolves as v is increased from 0 to v_{MI}.

(3.3) a) The bandstructure of the monovalent metal Cu is such that the highest occupied band lies completely below the Fermi energy in the vicinity of the $(\frac{1}{2}, \frac{1}{2}, \frac{1}{2})$ point of the Brillouin zone. Describe the form of the Fermi surface. (If stuck, you might find it helpful to look at Fig. 4.8.)

b) The magnitude of the energy gap at $(\frac{1}{2}, \frac{1}{2}, \frac{1}{2})$ is 4 eV and it is approximately independent of wavevector near this point. Estimate the radius of the Fermi surface 'neck'. (You will need the value of E_F calculated in Exercise 1.1.)

(3.4)* a) A monovalent two-dimensional metal crystallizes in a square lattice with lattice constant 0.3 nm. Experiments show that the Fermi surface of this material is extended (i.e. it crosses the Brillouin zone boundaries) along the set of $\langle 0\ 1 \rangle$ directions. Use simple nearly-free electron theory to explain how this can occur.

b) Measurements of the optical properties of this material show that there is a strong absorption feature in the visible region of the spectrum. Calculate the energy of this feature assuming that the Fermi surface only just touches the Brillouin-zone boundary.

(3.5) The condition for Bragg diffraction of a particle with wavelength λ is stated as

$$\lambda = 2d \sin \theta \qquad (3.24)$$

where d is the separation of the crystal planes involved and θ is the angle between the incident wavevector of the particle and the crystal planes (2θ is the angle between the incident and scattered wavevectors of the particle).[7]

Show that this is equivalent to

$$G^2 = 2\mathbf{k}.\mathbf{G}, \qquad (3.25)$$

where $|\mathbf{k}| = 2\pi/\lambda$. Choose a set of planes with Miller indices $(m_1, m_2 m_3)$, and let $\mathbf{G}$ have the same indices; if you have not already done so (see Execise 2.2) you will need to show that the spacing $d(m_1, m_2, m_3)$ of lattice planes with the Miller indices (m_1, m_2, m_3) (see Appendix A) is given by

$$d(m_1, m_2, m_3) = \frac{2\pi}{|\mathbf{G}(m_1, m_2, m_3)|},$$

where $\mathbf{G}(m_1, m_2, m_3) = m_1 \mathbf{A}_1 + m_2 \mathbf{A}_2 + m_3 \mathbf{A}_3$.

---

[7] See, for example, *Introduction to solid state physics*, by Charles Kittel, seventh edition (Wiley, New York, 1996) Chapter 2 for the conventional statement of the Bragg condition.

4 The tight-binding model

- 4.1 Introduction — 32
- 4.2 Band arising from a single electronic level — 32
- 4.3 General points about the formation of tight-binding bands — 35

4.1 Introduction

In the tight-binding model we assume the opposite limit to that used for the nearly-free-electron approach, i.e. the potential is so large that the electrons spend most of their lives bound to ionic cores, only occasionally summoning the quantum-mechanical wherewithal to jump from atom to atom.

The atomic wavefunctions $\phi_j(\mathbf{r})$ are defined by

$$\mathcal{H}_{\text{at}}\phi_j(\mathbf{r}) = E_j \phi_j(\mathbf{r}), \tag{4.1}$$

where $\mathcal{H}_{\text{at}}$ is the Hamiltonian of a single atom.

The assumptions of the model are

(1) close to each lattice point, the crystal Hamiltonian $\mathcal{H}$ can be approximated by $\mathcal{H}_{\text{at}}$;
(2) the bound levels of $\mathcal{H}_{\text{at}}$ are well localised, i.e. the $\phi_j(\mathbf{r})$ are very small one lattice spacing away, implying that
(3) $\phi_j(\mathbf{r})$ is quite a good approximation to a stationary state of the crystal, as will be $\phi_j(\mathbf{r} + \mathbf{T})$.

Reassured by the second and third assumptions, we make the Bloch functions $\psi_{j,\mathbf{k}}$ of the electrons in the crystal from linear combinations of the atomic wavefunctions

$$\psi_{j,\mathbf{k}} = \sum_{\mathbf{T}} a_{\mathbf{k},\mathbf{T}} \phi_j(\mathbf{r} + \mathbf{T}). \tag{4.2}$$

We now consider a simple case.

4.2 Band arising from a single electronic level

4.2.1 Electronic wavefunctions

We consider only the highest occupied orbital of an atom, $\phi(\mathbf{r})$, with energy E_ϕ. A linear combination of $\phi(\mathbf{r} - \mathbf{T})$ is used to produce a Bloch function $\psi_{\mathbf{k}}(\mathbf{r})$, which must obey Bloch's theorem (the following equation is merely an algebraic restatement of the discussion in Section 2.5)

$$\psi_{\mathbf{k}}(\mathbf{r} + \mathbf{T}) = e^{i\mathbf{k}\cdot\mathbf{T}} \psi_{\mathbf{k}}(\mathbf{r}). \tag{4.3}$$

These considerations (see Exercise 4.1) lead us to make up the tight-binding wavefunction

$$\psi_{\mathbf{k}}(\mathbf{r}) = \sum_{\mathbf{T}} e^{i\mathbf{k}\cdot\mathbf{T}} \phi(\mathbf{r} - \mathbf{T}). \tag{4.4}$$

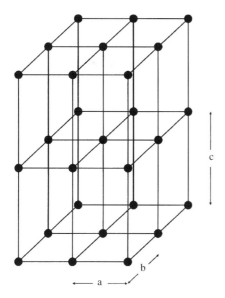

Fig. 4.1 Simple crystal structure (with orthogonal primitive lattice vectors) for tight-binding model derivation; the points indicate the positions of the atoms.

The $\psi_{\mathbf{k}}(\mathbf{r})$ satisfy the Bloch condition whilst displaying the atomic character of the levels.

4.2.2 Simple crystal structure.

In order to illustrate the role of the real-space positions of the atoms in determining the tight-binding bands, we consider a metal with atoms arranged to form a crystal with $\mathbf{a}_1 = a\mathbf{e}_1$, $\mathbf{a}_2 = b\mathbf{e}_2$ and $\mathbf{a}_3 = c\mathbf{e}_3$, where the $\mathbf{e}_j$ are the (mutually perpendicular) Cartesian unit vectors (see Fig. 4.1).

4.2.3 The potential and Hamiltonian

The potentials involved are

- the actual crystal potential $V(\mathbf{r})$;
- the potential associated with an isolated atom $V_0(\mathbf{r})$.

To first order, the crystal Hamiltonian is

$$\mathcal{H} = \mathcal{H}_{\text{at}} + \{V(\mathbf{r}) - V_0(\mathbf{r})\}, \tag{4.5}$$

with (of course) $\mathcal{H}\psi_{\mathbf{k}}(\mathbf{r}) = E(\mathbf{k})\psi_{\mathbf{k}}(\mathbf{r})$.

Operating with $\mathcal{H}$ on the expression for $\psi_{\mathbf{k}}(\mathbf{r})$ given in eqn 4.4 gives

$$\mathcal{H}_{\text{at}} \sum_{\mathbf{T}} e^{i\mathbf{k}.\mathbf{T}} \phi(\mathbf{r} - \mathbf{T}) + \{V(\mathbf{r}) - V_0(\mathbf{r})\} \sum_{\mathbf{T}} e^{i\mathbf{k}.\mathbf{T}} \phi(\mathbf{r} - \mathbf{T}) =$$

$$E(\mathbf{k}) \sum_{\mathbf{T}} e^{i\mathbf{k}.\mathbf{T}} \phi(\mathbf{r} - \mathbf{T}). \tag{4.6}$$

To evaluate $E(\mathbf{k})$ we multiply by $\phi^*(\mathbf{r})$ and integrate over all $\mathbf{r}$. Equation 4.6 becomes

$$\int \phi^*(\mathbf{r}) \mathcal{H}_{\text{at}} \sum_{\mathbf{T}} e^{i\mathbf{k}.\mathbf{T}} \phi(\mathbf{r} - \mathbf{T}) d^3\mathbf{r}$$

$$+ \int \phi^*(\mathbf{r}) \{V(\mathbf{r}) - V_0(\mathbf{r})\} \sum_{\mathbf{T}} e^{i\mathbf{k}.\mathbf{T}} \phi(\mathbf{r} - \mathbf{T}) d^3\mathbf{r}$$

$$= \int \phi^*(\mathbf{r}) E(\mathbf{k}) \sum_{\mathbf{T}} e^{i\mathbf{k}.\mathbf{T}} \phi(\mathbf{r} - \mathbf{T}) d^3\mathbf{r}. \quad (4.7)$$

The first integral in eqn 4.7 yields E_ϕ (the $\phi(\mathbf{r}-\mathbf{T})$ are normalised); the integral with $\mathbf{T} = 0$ is far greater than all of the others as $\phi(\mathbf{r})$ is so short range. The third integral in eqn 4.7 gives $E(\mathbf{k})$ for similar reasons. The second integral is the one which gives rise to the bands; it is non-negligible only for the on-site ($\mathbf{T} = 0$) and nearest-neighbour values of $\mathbf{T}$, i.e. $\mathbf{T} = \pm a\mathbf{e}_1, \pm b\mathbf{e}_2, \pm c\mathbf{e}_3$. The integral thus splits into four distict terms

$$\int \phi^*(\mathbf{r}) \{V(\mathbf{r}) - V_0(\mathbf{r})\} \sum_{\mathbf{T}} e^{i\mathbf{k}.\mathbf{T}} \phi(\mathbf{r} - \mathbf{T}) d^3\mathbf{r} =$$

$$\int \phi^*(\mathbf{r})(V - V_0) \phi(\mathbf{r}) d^3\mathbf{r}$$

$$+ (e^{ik_x a} + e^{-ik_x a}) \int \phi^*(\mathbf{r})(V - V_0) \phi(\mathbf{r} + \mathbf{a}_1) d^3\mathbf{r}$$

$$+ (e^{ik_y b} + e^{-ik_y b}) \int \phi^*(\mathbf{r})(V - V_0) \phi(\mathbf{r} + \mathbf{a}_2) d^3\mathbf{r}$$

$$+ (e^{ik_z c} + e^{-ik_z c}) \int \phi^*(\mathbf{r})(V - V_0) \phi(\mathbf{r} + \mathbf{a}_3) d^3\mathbf{r}. \quad (4.8)$$

Collecting all of this information together we obtain

$$E(\mathbf{k}) = E_\phi - B - 2t_x \cos(k_x a) - 2t_y \cos(k_y b) - 2t_z \cos(k_z c), \quad (4.9)$$

with

$$B = -\int \phi^*(\mathbf{r})(V - V_0) \phi(\mathbf{r}) d^3\mathbf{r} \quad (4.10)$$

$$t_x = -\int \phi^*(\mathbf{r})(V - V_0) \phi(\mathbf{r} + \mathbf{a}_1) d^3\mathbf{r} \quad (4.11)$$

$$t_y = -\int \phi^*(\mathbf{r})(V - V_0) \phi(\mathbf{r} + \mathbf{a}_2) d^3\mathbf{r} \quad (4.12)$$

$$t_z = -\int \phi^*(\mathbf{r})(V - V_0) \phi(\mathbf{r} + \mathbf{a}_3) d^3\mathbf{r}. \quad (4.13)$$

The parameters t_x, t_y and t_z are known as **transfer integrals**. They give an indication of how easy it is for an electron to transfer from one atom to the next.

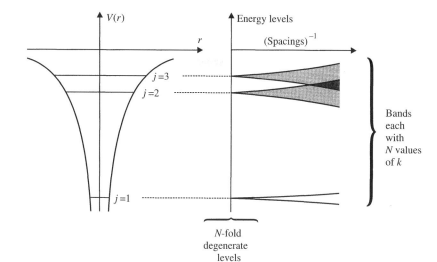

Fig. 4.2 Schematic representation of the formation of tight-binding bands as the spacing between atoms is reduced.

4.3 General points about the formation of tight-binding bands

The derivation in the previous section illustrates several points about real bandstructure.

(1) Fig. 4.2 shows schematically the process involved in forming the tight-binding bands. N single atoms with j (doubly degenerate) atomic levels have become j $2N$-fold degenerate bands.

(2) The transfer integrals give a direct measure of the width in energy of a band (the **bandwidth**); small transfer integrals will give a narrow bandwidth. Exercises 5.1 and 5.5 show that carriers close to the bottom of a tight-binding band have **effective masses** which are inversely proportional to the transfer integrals. As will be seen in Chapter 5, the effective mass parameterises the ease with which an electron can be accelerated. Thus, small transfer integrals lead to heavy effective masses (and narrow bandwidths), indicating that it is hard to move the electrons around; large transfer integrals lead to light effective masses (and large bandwidths), indicating that it is easy to move the electrons around. This ties in exactly with the real-space picture of the formation of tight-binding bands, in which the transfer integrals reflect the ease with which an electron can transfer from atom to atom.

(3) The 'shape' of the bands in k-space will be determined in part by the real-space crystal structure; if the atoms in a certain direction are far apart, then the bandwidth will be narrow for motion in that direction.

(4) The bands will also reflect the character of the atomic levels which have gone to make them up.

The latter point will be illustrated in Section 4.3.3 below.

This is more-or-less all that one needs to know to understand most of the properties of real bands. The reader in a hurry may now rush straight on to Chapter 5 and the rest of the book without pausing, if she or he wishes. Those of a more meandering nature may persist with this chapter a little longer for a qualitative explanation of the electronic properties of some of the elements, couched in terms of the atomic orbitals from which their bands are derived.

4.3.1 The group IA and IIA metals; the tight-binding model viewpoint

Figure 4.3 shows a schematic of the formation of tight-binding bands in a metal from Group IA of the periodic table, such as Na (left-hand side), and a Group IIA metal, such as Ca (right-hand side); the vertical dotted lines show a typical atomic spacing. The lower orbitals, such as 1s, are small in spatial extent, so that the overlap of wavefunctions on neighbouring atoms will be small, giving a small bandwidth; such bands will tend to be well-separated in energy. However, the higher-energy orbitals will be more voluminous, so that the overlap of wavefunctions will increase, leading to a larger bandwidth. Taking Na as an example, the 1s, 2s and 2p orbitals will form narrow bands that are completely full up; the 2p band is the lowest band shown in the schematic of Fig. 4.3. The 3s band will be wider, and half-filled, as Na is monovalent; each atom provides one 3s electron. Hence, Na will be a metal, as empty electron states are immediately adjacent in energy to filled states.

In the schematic of Ca (Fig. 4.3, right-hand side), the atoms are sufficiently close together for the bands derived from the 3s and 3p orbitals to overlap, so that the element is also a metal.

As one descends groups IA and IIA of the periodic table, the same sort of considerations will tend to apply. The orbitals of the ionic core will have little overlap, and give rise to energy bands that are very narrow and completely filled; virtually all of the electronic properties of the metal will be determined by the outermost s and possibly p orbitals, which will give rise to a partially filled band or bands. Hence, the group IA metals (the alkali metals; Li, Na, K etc.) are rather similar to each other from an electronic point-of-view. Likewise, the alkaline earths (the group IIA metals Be, Mg, Ca, Sr, Ba) all have quite similar properties.

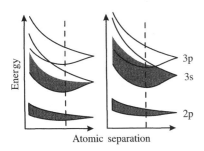

Fig. 4.3 Schematic of the formation of tight-binding bands in a group IA metal (left-hand picture) and a group IIA metal (right-hand picture). The dotted line represents a typical atomic spacing in a metal; the shading represents filled states within the bands. Note that I have also shown the curving up of the band energies which occurs at small values of atomic spacing; those unsure as to why this occurs should see Fig. 2.6 of *Semiconductor physics*, by K. Seeger (Springer, Berlin, 1991) for an explanation.

4.3.2 The Group IV elements

The elements C (Diamond), Si, Ge and Sn all crystallise in what is known as the **diamond structure**; it can be visualised as two interlocking face-centred-cubic lattices (i.e. as a face-centred cubic lattice with a two-atom basis — see Fig. 4.4). In this structure, each atom has tetragonal symmetry, with four bonds to its nearest neighbours. Taking carbon ($2s^2\ 2p^2$) as an example, the atoms form bonding and antibonding states for both s and p orbitals (see Fig. 4.5)[1]; the lowest four states from each atom mix to form sp^3 hybrids. As there are two atoms per primitive cell, each with four electrons, the lower sp^3 states, which overlap to form a band, are completely filled. Thus, diamond is not metallic, as the occupied and unoccupied electron states are separated by a considerable energy gap.

It is interesting to examine what happens as one descends group IV of the periodic table. Table 4.1 shows the energy gap separating the filled sp^3 hybrid band from the unoccupied band derived from the antibonding orbitals and the

[1] The mechanisms leading to the formation of bonding and antibonding orbitals in molecules and assemblies of atoms are not strictly relevant to the qualitative discussion in this section; details are given in most quantum mechanics or atomic physics texts, e.g. Chapters 20 and 21 of *Quantum mechanics*, by Stephen Gasiorowicz (Wiley, New York, 1974); see also Fig. 2.6 of *Semiconductor physics*, by K. Seeger (Springer, Berlin, 1991).

4.3 General points about the formation of tight-binding bands

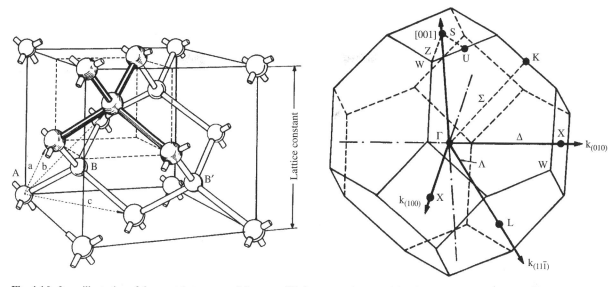

Fig. 4.4 Left: an illustration of the crystal structures of the group IV elements and many of the binary semiconductors such as GaAs and CdTe. In the case of C (diamond), Si, Ge and Sn, the atoms on the A and B sites are identical (the so-called diamond structure). In the case of binary semiconductors such as for example GaAs, the A will be Ga and B As; the crystal is said to have the zinc blende structure. In each case, the underlying lattice is face-centred cubic and there is a two-atom basis. Thus, both the group IV elements and the binary semiconductors have the same shape of first Brillouin zone, which is shown on the right.

parameter a, the edge of the conventional cubic cell shown in Fig. 4.4; a is a measure of the atomic separation. Note that the band gap decreases as the separation of the atoms increases, until Sn is actually metallic.

The reason for this variation is illustrated in Fig. 4.6, which shows the variation of the tight-binding bands with interatomic distance. The larger atomic spacing in Sn corresponds to the portion of the diagram where the upper and lower bands overlap, so that Sn is a metal; as the atomic spacing decreases, the gap between the band derived from the sp^3 bonding states and that derived from the antibonding states opens up, resulting in an energy gap between full and empty states.

4.3.3 The transition metals

I have already mentioned above that the tight-binding model implies that bands will reflect the character of the atomic levels which have gone to make them up. This is illustrated by Fig. 4.7, which shows the calculated tight-binding bands for copper. Note that the character of the original atomic levels is reflected in the width and shape of the the bands; the more compact and anisotropic 3d orbitals give rise to five narrow bands of complex shape, whilst the single band derived from the larger, spherical 4s orbitals is wide and almost free-electron-like. The properties of a particular 3d metal are strongly dependent on the position of the Fermi energy amongst this mess.

Note that the colours of 3d metals such as Cu and Au result from the optical transitions which are possible between the occupied d-bands and the empty states at the top of the s-band.

Figure 4.7 also shows that the band derived from the 4s orbitals in Cu dips

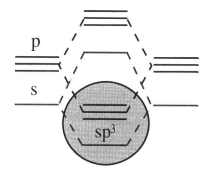

Fig. 4.5 Schematic of the formation of sp^3 hybrid bonding states in carbon (diamond).

Table 4.1 Energy gaps and lattice parameters a for the Group IV elements; a is the edge of the conventional cubic cell shown in Fig. 4.4. (The band-gaps shown are the direct gaps; see Chapter 6.)

Element	E_g	a
C	5 eV	0.356 nm
Si	1.1 eV	0.543 nm
Ge	1.0 eV	0.566 nm
Sn	metallic	0.646 nm

38 *The tight-binding model*

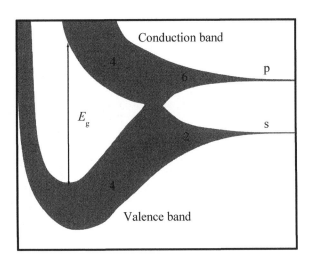

Fig. 4.6 Schematic of tight-binding band formation in the group IV elements; the electron energy is plotted against interatomic spacing. The larger atomic spacing in Sn corresponds to the portion where the upper and lower bands overlap, so that Sn is a metal; as the atomic spacing decreases, the gap between the band derived from the sp^3 bonding states and that derived from the antibonding states opens up, resulting in an energy gap between full and empty states.

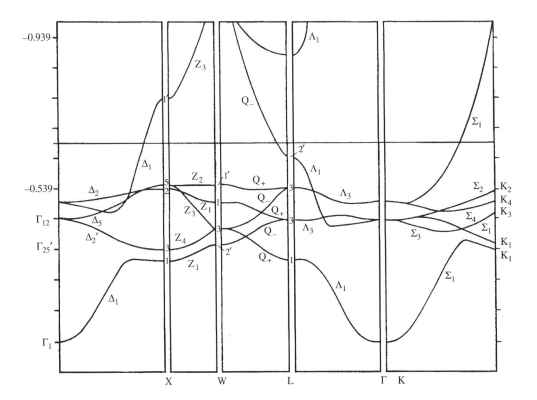

Fig. 4.7 Theoretical calculation of the bandstructure of copper. The electron energy (in units of Rydbergs, i.e. 13.6 eV) is plotted as a function of **k** for various directions in the Brillouin zone (see Fig. 4.8 for the notation used). Note that the band derived from the 4s orbitals (starting at Γ_1, it anticrosses with the bands derived from the 3d orbitals, before re-emerging as Δ_1) is very similar to the free-electron dispersion relationship. The Fermi energy is indicated for Cu; the d-bands are deeply buried, and Cu in many respects looks reasonably like a metal obeying the Sommerfeld model; apart from the necks (see Fig. 4.8), the Fermi surface is quite spherical. On the other hand, in the case of a metal such as Co, the Fermi energy is in amongst the d-bands, resulting in a complex Fermi surface and very different properties. (Figure taken from G.A. Burdick, *Phys. Rev.* **129**, 138 (1963).)

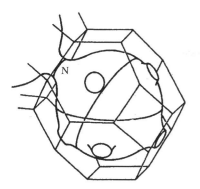 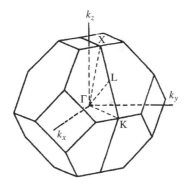

Fig. 4.8 Fermi surface and Brillouin zone of Cu, illustrating the famous 'necks' (labelled 'N') at the L points on the Brillouin-zone boundaries.

below E_F at the L-points on the Brillouin-zone boundaries. This results in the famous Fermi surface with its *necks*, shown in Fig. 4.8.[2]

Further reading

Alternative treatments are given in *Introduction to solid state physics*, by Charles Kittel, seventh edition (Wiley, New York, 1996) Chapter 9 (simpler), *Solid state physics*, by G. Burns (Academic Press, Boston, 1995) Section 10.9 (see also Section 10.3) (simpler) and *Solid state physics*, by N.W Ashcroft and N.D. Mermin (Holt, Rinehart and Winston, New York, 1976) pages 176–190 (more general and therefore more complicated). The tight-binding model is related to the **Kronig–Penney model**, which is a very simple illustration of the formation of bands; see e.g. *Quantum mechanics*, by Stephen Gasiorowicz (Wiley, New York, 1974) page 98.

More advanced methods for calculating bandstructures of solids are reviewed in *Quantum theory of the solid state*, by Joseph Callaway (Academic Press, San Diego, 1991), Chapter 1.

[2] The geometry of the Brillouin zone and size of Fermi surface are slightly less favourable for free-electron-like behaviour in the case of Cu than they are for Na or K, because the free-electron Fermi surface comes closer to the zone edge in the former case. Cu has a face-centred-cubic lattice; if we just consider the single electron derived from the 4s orbital, then the free-electron Fermi wavevector is $k_F = 4.90/a$, with a the conventional unit cell cube edge. The shortest dimension of the Brillouin zone is $|\mathbf{A}_j|/2 = \sqrt{3}\pi/a$; the ratio $k_F/(|\mathbf{A}_j|/2)$ is therefore 0.90, rather than the 0.88 of the body-centred cubic Na or K.

Exercises

(4.1) The tight-binding wavefunction proposed in eqn 4.4 is

$$\psi_\mathbf{k}(\mathbf{r}) = \sum_\mathbf{T} e^{i\mathbf{k}.\mathbf{T}} \phi(\mathbf{r} - \mathbf{T}). \quad (4.14)$$

Show that $\psi_\mathbf{k}(\mathbf{r})$ obeys Bloch's law

$$\psi_\mathbf{k}(\mathbf{r} + \mathbf{T}) = e^{i\mathbf{k}.\mathbf{T}} \psi_\mathbf{k}(\mathbf{r}). \quad (4.15)$$

(4.2) In this question, we examine the various Fermi-surface shapes which can occur in a crystal with the primitive lattice translation vectors $\mathbf{a}_1 = a\mathbf{e}_1$, $\mathbf{a}_2 = b\mathbf{e}_2$ and $\mathbf{a}_3 = c\mathbf{e}_3$ which possesses the tight-binding band.

$$E(\mathbf{k}) = E_0 - 2t_x \cos(k_x a) - 2t_y \cos(k_y b) - 2t_z \cos(k_z c). \quad (4.16)$$

a) **Two-dimensional tight-binding model.** Let us assume for the moment that the crystal is two-dimensional (i.e. set $t_z = 0$ and forget about the z-direction) and that $a = b$ and $t_x = t_y$. Sketch the Fermi surface in the k_x, k_y plane when the band is almost empty (i.e. when there is $\ll 1$ electron per primitive unit cell) and when the band is almost full (i.e. when there are just less than two electrons per primitive unit cell). Show that when the band is exactly half filled (i.e. exactly one electron is donated by each unit cell) the Fermi surface is a square that just touches the Brillouin zone boundaries. Sketch also the Fermi surface in the case that the band is half filled but $a \approx b$ and $t_x \gg t_y$. (This is quite close to shape of the Fermi surfaces of many organic conductors.)

b) **Three-dimensional tight-binding model.** Sketch the Fermi surfaces for almost empty and almost full bands for $a = b \neq c$, $t_x = t_y \gg t_z$. It is most rewarding to sketch these in the k_x, k_z plane using the extended zone scheme.

Some general points about bandstructure

5.1 Comparison of tight-binding and nearly-free-electron bandstructure

5.1	Comparison of tight-binding and nearly-free-electron bandstructure	41
5.2	The importance of k	42
5.3	Holes	45
5.4	Postscript	46

Let us compare the lowest band of the nearly-free-electron model in Fig. 3.3 with a one-dimensional tight-binding band (cf. Equation 4.9)

$$E(k) = E_0 - 2t\cos(ka), \qquad (5.1)$$

where E_0 is a constant. Note that

- both bands look qualitatively similar, i.e.
- both bands have minima and maxima (i.e. points where $dE/dk = 0$) at the Brillouin-zone centres and boundaries respectively;
- both bands have the same k-space periodicity;
- the tops and bottoms of both bands are approximately parabolic (i.e. E is proportional to $(\mathbf{k} - \mathbf{k}_0)^2$ plus a constant, where $\mathbf{k}_0$ is the point in k-space at which the band extremum occurs).

The last point is obvious for the bottom of the lowest nearly-free-electron band; it is proved for the highest points of this band in Exercise 5.4. To see that the same thing happens for the tight-binding approach, we use the approximation

$$\cos(ka) \approx 1 - \frac{(ka)^2}{2} \qquad (5.2)$$

for small ka in eqn 5.1 to give

$$E(k) \approx E_0 - 2t + ta^2k^2. \qquad (5.3)$$

The equivalent proof for the top of the band (close to π/a) is left to the reader (see also Exercises 5.1 and 5.5).

The above discussion shows that both very weak and very strong periodic potentials give rise to qualitatively similar bands. We therefore expect that our ideas about bands derived from these simple models will apply to the real bands in solids, where the potential strengths will fall between the two extremes. The more complex arrangements of atoms and/or molecules often found in real solids will give rise to more complex bandshapes, but qualitatively the properties of the bands will be the same as those of our simple models.

5.2 The importance of k

5.2.1 $\hbar \mathbf{k}$ is *not* the momentum

Bloch's theorem has introduced wavevectors which we have labelled $\mathbf{k}$ and $\mathbf{q}$ (see e.g. eqn 2.25). Although at first sight $\mathbf{k}$ and $\mathbf{q}$ look similar to the straightforward electron momentum $\mathbf{p} = \hbar \mathbf{k}$ in the Sommerfeld model, it is easy to show that the Bloch functions are not eigenstates of the momentum operator;

$$\frac{\hbar}{i}\nabla \psi(\mathbf{r}) = \frac{\hbar}{i}\nabla e^{i\mathbf{q}\cdot\mathbf{r}} u_{j,\mathbf{q}} = \hbar \mathbf{q}\psi(\mathbf{r}) + e^{i\mathbf{q}\cdot\mathbf{r}}\frac{\hbar}{i}\nabla u_{j,\mathbf{q}} \neq \mathbf{p}\psi(\mathbf{r}) \quad (5.4)$$

(where the Bloch wavefunction used is defined in eqn 2.25).

The quantity $\hbar \mathbf{k}$ is instead a **crystal momentum**, the momentum of the system as a whole; it is better to think of $\mathbf{k}$ as a quantum number which describes a Bloch state.

5.2.2 Group velocity

In the original formulation of the Bloch states (see eqn 2.16), the electron wavefunction was defined to be the superposition of a set of plane waves (a *wavepacket*). It is therefore possible to use the idea of a *group velocity* $\mathbf{v}$

$$\mathbf{v} = \frac{1}{\hbar}\nabla_\mathbf{k} E, \quad (5.5)$$

where $\nabla_\mathbf{k}$ is the gradient operator in k-space, to describe the real-space motion of the electron (cf. $v = (d\omega/dk)$) for the wavepackets encountered in optics). Bearing this in mind, we are now going to derive the **effective mass**, which is very useful in parameterising the dynamics of band electrons when they are subjected to external forces. The derivation also illustrates a very important point about $\mathbf{k}$. I shall follow the derivation through in one dimension; Exercise 5.2 involves doing the same thing in three dimensions.

5.2.3 The effective mass

Let an external force f be applied to a band electron. The force will do work

$$\delta E = fv\delta t \quad (5.6)$$

in time δt. In addition,

$$\delta E = \frac{dE}{dk}\delta k = \hbar v \delta k. \quad (5.7)$$

Equating eqns 5.6 and 5.7, dividing through by δt and considering the limit $\delta t \to 0$ gives

$$\hbar \frac{dk}{dt} = f. \quad (5.8)$$

The equivalent three-dimensional formula in the derivation of Exercise 5.2, is

$$\hbar \frac{d\mathbf{k}}{dt} = \mathbf{f}, \quad (5.9)$$

where **k** and **f** are now vectors. Equation 5.9 is enormously important; it shows that *in a crystal* $\hbar \frac{d\mathbf{k}}{dt}$ *is equal to the external force on the electron.*

After this amazing fact, the rest of the effective mass derivation is almost an anticlimax. The rate of change of velocity with time is

$$\frac{dv}{dt} = \frac{1}{\hbar}\frac{d^2E}{dk\,dt} = \frac{1}{\hbar}\frac{d^2E}{dk^2}\frac{dk}{dt}. \tag{5.10}$$

Substituting for $\frac{dk}{dt}$ from eqn 5.8 and rearranging gives

$$\frac{\hbar^2}{\frac{d^2E}{dk^2}}\frac{dv}{dt} = m^*\frac{dv}{dt} = f, \tag{5.11}$$

where the *effective mass* m^* is defined by

$$m^* = \frac{\hbar^2}{\frac{d^2E}{dk^2}}. \tag{5.12}$$

(The equivalent tensor quantities for an anisotropic band are given in Exercise 5.2.)

Equation 5.11 shows that the effective mass gives a convenient way of describing the motion of band electrons subjected to an external force; the identity 'force = (mass × acceleration)' can be used with the effective mass substituted for the inertial mass.

In general, the effective mass will be energy-dependent. However, one often has to deal with almost empty or almost full bands, i.e. the states close to the minima and maxima of the $E_j(\mathbf{k})$ dispersion relationships. As we have already seen in Section 5.1, these regions are often approximately parabolic (this is actually a general property of any maximum or minimum– try expanding any function possessing such an extremum close to that point using a Taylor series). Close to these points, the electrons can therefore be treated as if they were free, but with an effective mass; i.e. their dispersion relationship is

$$E(\mathbf{k}) \approx E_0 + \frac{\hbar^2}{2m^*}(\mathbf{k} - \mathbf{k_0})^2, \tag{5.13}$$

where energy E_0 and wavevector $\mathbf{k_0}$ define the band extremum. Note that m^* can be either positive or negative. This approximation often simplifies calculations; m^* in effect contains all of the necessary information about the way in which the electron's motion is modified by the crystal potential through which it moves. The equivalent model for anisotropic band extrema is explored in Exercise 5.2.

5.2.4 The effective mass and the density of states

In the previous chapters, we have seen that it is most natural to count electron states by evaluating the volume of k-space occupied (e.g. the Brillouin-zone volume determines the number of electrons that a band can accommodate). The counting of states is easy, because the states are spread through phase-space uniformly. However, when one comes to consider the evaluation of quantities such as the electronic heat capacity, where thermal population of states is a significant factor, it is often convenient to work in terms of energy (see e.g. the derivation of the electronic heat capacity within the Sommerfeld model in Chapter 2), requiring a knowledge of the number of states per unit energy range per unit volume, $g(E)$.

If one substitutes a general value of k in eqn 1.22 instead of k_F, one obtains $n(\mathbf{k})$, the number of states per unit volume of r-space with wavevectors less than $|\mathbf{k}|$. If the band in question is assumed to obey eqn 5.13 close to the region of interest (i.e. it is parabolic and isotropic), then

$$g(E) \equiv \frac{dn}{dE} = \frac{dn}{dk}\frac{dk}{dE} = \frac{1}{2\pi^2}\left(\frac{2m^*}{\hbar^2}\right)^{\frac{3}{2}}(E - E_0)^{\frac{1}{2}}. \quad (5.14)$$

This shows that the effective mass is a very convenient way of parameterising the curvature of a band at a certain energy, and hence the density of states $g(E)$ at that energy (see Exercise 1.1). We shall use this technique in a number of subsequent derivations.

Finally, note that eqn 5.14 shows that a heavy effective mass results in a large density of states (see also Exercise 1.1). The reason for this can be seen in Fig. 5.1, which shows a parabolic band with a light effective mass (steep curve) compared to a band with a heavy effective mass (shallow curve). A fixed region of k-space will always contain a fixed number of states. In the case of the heavy-mass band, the k-space interval Δk corresponds to a small interval of energy δE_1; the same interval of k corresponds to a much larger energy interval δE_2 for the light-mass band. As the same number of states is accommodated in each energy region, the density of states $g(E)$ is much higher for the heavy-mass band.

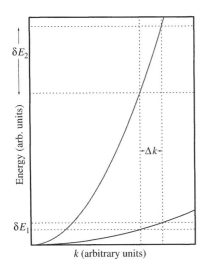

Fig. 5.1 Comparison of a parabolic band with a light effective mass (steep curve) with a band with a heavy effective mass (shallow curve). In the case of the heavy-mass band, the k-space interval Δk corresponds to a small interval of energy δE_1; the same interval of k corresponds to a much larger energy interval δE_2 for the light-mass band.

5.2.5 Summary of the properties of k

The properties of the **k** used in the Bloch wavefunction (see eqns 2.16 and 2.25) may be summarised as follows.

(1) **k** is not the (physical momentum/$\hbar$) of the electron; it is a quantum number describing the electron's state within a band. Each band is labelled using the index j.

(2) For each j, **k** takes all values consistent with the Born–von Karman boundary conditions within the first Brillouin zone; in our notation **k** runs though all **q**. j can run through an infinite number of discrete values.

(3) For a given j, the electronic dispersion relationship $E_j(\mathbf{k})$ has no explicit form; the only constraint is that it must be periodic, i.e. $E_j(\mathbf{k}) = E_j(\mathbf{k}+\mathbf{G})$.

(4) The velocity of an electron with energy $E_j(\mathbf{k})$ is given by $\mathbf{v} = \frac{1}{\hbar}\nabla_\mathbf{k} E_j(\mathbf{k})$.

(5) The rate of change of **k** under the action of an external force **f** is given by $\hbar \frac{d\mathbf{k}}{dt} = \mathbf{f}$.

(6) The electronic wavefunction is $\psi(\mathbf{r}) = e^{i\mathbf{k}\cdot\mathbf{r}} u_{j,\mathbf{k}}$, where $u_{j,\mathbf{k}}(\mathbf{r}) = u_{j,\mathbf{k}}(\mathbf{r}+\mathbf{T})$. Here $u_{j,\mathbf{k}}(\mathbf{r})$ has no simple explicit form.

The above list tells us all we need to know about the dynamics of electrons in solids. Let us imagine for a line or two that we know nothing about bands; a calculation of the motion of an electron in a solid due to an external force seems intractable, as the electron is also subject to a complex array of forces from the ionic cores/molecules in the crystal. Band theory gets round this problem very effectively. The external force results in a time-dependent quantum number $\mathbf{k}(t)$. The resulting electronic velocity is obtained from the bandstructure using $\mathbf{v} = (1/\hbar)\nabla_\mathbf{k} E_j(\mathbf{k})$ and substituting $\mathbf{k} = \mathbf{k}(t)$. The bandstructure thus tells us all that we need to know about the interaction between the electrons and the relatively static components of the crystal (ions, molecules etc.). Exercise 5.1 is a good example of this technique.

5.2.6 Scattering in the Bloch approach

The Bloch wavefunctions describe states which are stationary solutions to the Schrödinger equation of a periodic potential. *Thus these states persist for ever and ever in a perfectly periodic infinite crystal*; there will be no scattering of electrons in the absence of disorder (which disturbs the perfect periodicity) or boundaries (which destroy the periodicity/infiniteness of the crystal). This is a manifestation of the wave-like nature of electrons; in a periodic array of scatterers a wave can propogate without attenuation because of the coherent constructive interference of the scattered waves. Scattering only takes place because of lack of periodicity in r-space.

This observation is in great contrast with the Drude model (where electrons were pictured crashing into virtually every ion) and the Sommerfeld approach (where scattering was acknowledged but its cause was a mystery). In addition

- it encompasses all possible causes of scattering (phonons (see Appendix D), other electrons, impurities and boundaries, all of which disturb the local periodicity);
- it explains the enormous low-temperature scattering lengths observed in very pure single crystals of metals (often $\sim$ mm or even cm);
- it removes the need for the hand-waving explanations as to why the electrons in a metal 'might not see the ionic cores very much' used to justify the Drude and Sommerfeld approaches.

5.3 Holes

It is going to be useful to be able to describe a few empty states close to the top of an almost full band using the concept of **holes**. Consider a band, containing electrons with quantum numbers $\mathbf{k}_j$, velocities $\mathbf{v}_j$ and energies $E(\mathbf{k}_j)$, where $E = 0$ is at the top of the band. For a full band, the values of $\mathbf{k}$ should all sum to zero, i.e.

$$\sum_j \mathbf{k}_j = 0. \tag{5.15}$$

We consider removing one electron to create an excitation which we label a hole. Suppose that the lth electron is removed; the band then acquires a net $\mathbf{k}$ which we attribute to the presence of the hole. The hole has $\mathbf{k} = \mathbf{k}_\mathrm{h}$, with

$$\mathbf{k}_\mathrm{h} = \sum_{j \neq l} \mathbf{k}_j = -\mathbf{k}_l. \tag{5.16}$$

It is obvious that the lower down the band the empty state, the more excited the system. The hole's energy E_h must therefore take the form

$$E_h = -E(\mathbf{k}_l). \tag{5.17}$$

The group velocity $\mathbf{v}_h$ associated with the hole is

$$\mathbf{v}_h = \frac{1}{\hbar} \nabla_{\mathbf{k}_h} E_h = \frac{1}{\hbar} \nabla_{-\mathbf{k}_l}(-E(\mathbf{k}_l)) = \mathbf{v}_l, \tag{5.18}$$

where the minus signs in eqns 5.16 and 5.17 have cancelled each other.

The full band will carry no current, i.e.

$$\sum_j (-e)\mathbf{v}_j = 0; \tag{5.19}$$

the removal of the lth electron produces a current

$$\sum_{j \neq l} (-e)\mathbf{v}_j = -(-e)\mathbf{v}_l = (+e)\mathbf{v}_h, \tag{5.20}$$

i.e. the hole appears to have an associated *positive* charge. Finally, substitution of eqns 5.16 and 5.17 into eqn 5.12 shows that the effective mass m_h^* (or effective mass tensor components in a more complex band; see Exercise 5.2) associated with the hole is given by

$$m_h^* = -m_l^*. \tag{5.21}$$

The importance of holes stems from the fact that bands are often well-characterised only close to the band extrema. Our knowledge of the dispersion relationships away from a particular extremum is often rather nebulous; such regions are not easily studied experimentally (see Chapter 8). It is therefore easier to deal with a small number of empty states close to the well-characterised maximum of an almost full band rather than attempt to treat the huge number of poorly-characterised states lower down in the band.

The fact that holes behave as though they have positive charge and positive effective mass explains why many di- and trivalent metals have positive Hall coefficients (see Section 1.4.3 and Table 1.1). A simple instance in which this could happen is shown in Fig. 3.4. The section of Fermi surface in the corner of the Brillouin zone (Fig. 3.4(e)) corresponds to a small number of empty states at the top of a band; the states are therefore hole-like. By contrast, the sections of Fermi surface straddling the zone boundaries (Fig. 3.4(f)) represent a few filled states at the bottom of the upper band; these states are therefore electron-like. The observed Hall coefficient would depend on the relative contributions that the various Fermi-surface sections make to the electrical conductivity (see Section 10.2).

5.4 Postscript

We have now derived all of the ideas needed for a reasonable understanding of the bandstructures of real solids. We shall take these ideas and use them to study the properties of semiconductors and insulators in the following chapter.

Further reading

A simple treatment is found in, for example, *Electricity and magnetism*, by B.I. Bleaney and B. Bleaney, third edition (Oxford University Press, Oxford, 1989), Sections 12.4 and 12.5, and a more comprehensive version is available in *Solid state physics*, by N.W Ashcroft and N.D. Mermin (Holt, Rinehart and Winston, New York, 1976), Chapter 12.

A higher-level description of the general properties of several different types of bandstructure calculation is given in *Quantum theory of the solid state*, by Joseph Callaway (Academic Press, San Diego, 1991), Chapter 1, *Semiconductor physics*, by K. Seeger (Springer, Berlin, 1991), Chapter 2, *Electrons in metals*, by J.M. Ziman (Taylor and Francis) and (harder) *Fundamentals of semiconductors*, by P. Yu and M. Cardona (Springer, Berlin, 1996), Chapter 2.

Exercises

(5.1) A one-dimensional solid, lattice spacing a, lies along the x-axis. It has a band with the following dispersion relationship

$$E(k) = E_0 - 2I\cos(ka). \quad (5.22)$$

The band contains just one electron which is at rest at $x = 0$ at time $t < 0$. At $t = 0$ an electric field ϵ is turned on in the x direction. Sketch the electron's position x, velocity v and effective mass m^* as a function of time t and comment on the relationship between the three quantities.

(5.2) **Reciprocal mass tensor and effective mass tensor.** The effective masses of electrons in anisotropic bands can be represented using tensors. This question exposes you to them.

a) Show that an electron in a crystal, with wavevector $\mathbf{k}$ and energy $E(\mathbf{k})$ subjected to an applied electric field ϵ, will change its velocity $\mathbf{v}$ at a rate given by

$$\frac{dv_i}{dt} = -e\sum_j (\mathbf{M}^{-1})_{ij}\epsilon_j \quad (5.23)$$

where the reciprocal mass tensor has components

$$(\mathbf{M}^{-1})_{ij} = \frac{1}{\hbar^2}\frac{\partial^2 E}{\partial k_i \partial k_j}. \quad (5.24)$$

b) The highest occupied band of a particular substance is found to have the following reciprocal mass tensor

$$\mathbf{M}^{-1} = \begin{pmatrix} \frac{1}{m_1} & 0 & 0 \\ 0 & \frac{1}{m_2} & 0 \\ 0 & 0 & \frac{1}{m_3} \end{pmatrix} \quad (5.25)$$

when a particular set of x, y and z axes are chosen. Sketch a constant energy surface (i.e. Fermi surface) of this substance, assuming that m_1, m_2 and m_3 are independent of E and $\mathbf{k}$ and that they differ in size.

c) As $\mathbf{M}^{-1}$ is diagonal, it is easy to invert it to get the effective mass tensor $\mathbf{M}$.

$$\mathbf{M} = \begin{pmatrix} m_1 & 0 & 0 \\ 0 & m_2 & 0 \\ 0 & 0 & m_3 \end{pmatrix}. \quad (5.26)$$

A magnetic field $\mathbf{B}$ is applied to the sample in an arbitrary direction. By substituting $\mathbf{v} = \mathbf{v_0}e^{i\omega t}$ into the equation

$$\mathbf{M}\frac{d\mathbf{v}}{dt} = -e\mathbf{v} \times \mathbf{B} \quad (5.27)$$

show that $\omega = \frac{eB}{m^*}$, with

$$m^* = \left(\frac{B^2 m_1 m_2 m_3}{m_1 B_x^2 + m_2 B_y^2 + m_3 B_z^2}\right)^{\frac{1}{2}}. \quad (5.28)$$

(Hint: write out all of the components of the equation and collect them into a matrix of coefficients of v_x, v_y and v_z; take a determinant of this and equate it to zero and everything should be all right.) This phenomenon is known as cyclotron resonance.

48 Some general points about bandstructure

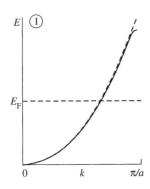

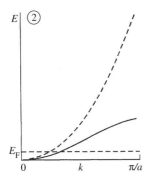

Fig. 5.2 Simplified E vs k curves for conduction electrons in Cu and Sc. The dashed lines indicate the Fermi energy in each case and the dotted curves represent the free-electron dispersion relationship.

(5.3) a) Figure 5.2 shows simplified electron dispersion relationships for (1) the s-electron conduction band of Cu ($4s^1$) and (2) the d-electron conduction band of Sc ($3d^1$). The dispersion relationship for free electrons is shown as a dotted line. Discuss the variation in magnitude and sign of the effective mass of electron states (both occupied and unoccupied) in these bands.
b) The Fermi energies of Cu and Sc are given as dashed lines in Fig. 5.2. Estimate the ratio of the low-temperature electronic molar heat capacities of Cu and Sc, $C_{el}(Cu)/C_{el}(Sc)$, stating clearly any assumptions that you make. What else can be deduced about the comparitive properties of Cu and Sc from Fig. 5.2?

(5.4)* We have proved that close to a Brillouin zone boundary, an electron in a **weak**, periodic potential has the following energy

$$E = \frac{1}{2}(E^0_{\mathbf{k}} + E^0_{\mathbf{k-G}}) \pm [\left(\frac{E^0_{\mathbf{k}} - E^0_{\mathbf{k-G}}}{2}\right)^2 + |V_{\mathbf{G}}|^2]^{\frac{1}{2}},$$

(5.29)

where $E^0_{\mathbf{k}} = \frac{\hbar^2 k^2}{2m_e}$ is the free-electron energy corresponding to wavevector $\mathbf{k}$, $\mathbf{G}$ is a reciprocal lattice vector, $V_{\mathbf{G}}$ is the appropriate coefficient from

$$V(\mathbf{r}) = \sum_{\mathbf{G}} V_{\mathbf{G}} e^{i\mathbf{G}\cdot\mathbf{r}},$$

(5.30)

and $V(\mathbf{r})$ is the potential. Consider a simple cubic lattice, with primitive lattice translation vectors $a\mathbf{e}_1$, $a\mathbf{e}_2$ and $a\mathbf{e}_3$, where $\mathbf{e}_1$, $\mathbf{e}_2$ and $\mathbf{e}_3$ are the (mutually perpendicular) Cartesian unit vectors parallel to the x, y and z axes respectively. Show that the effective masses for motion parallel to $\mathbf{e}_1$ with $\mathbf{k}$ close to the point $(\frac{\pi}{a}, 0, 0)$, are smaller than m_e and directly proportional to $|V_{\mathbf{G}}|$. Why do you obtain two masses, and what does their relative size tell you about the bands close to the zone boundary? Show also that the effective mass for motion parallel to $\mathbf{e}_2$ with $\mathbf{k}$ close to $(\frac{\pi}{a}, 0, 0)$ is m_e.

(5.5) A solid possesses the tight-binding band

$$E(\mathbf{k}) = E_0 - 2t_1 \cos(k_x a) - 2t_1 \cos(k_y b).$$ (5.31)

Evaluate the effective mass for motion in the k_x, k_y plane when the band is almost empty (i.e. when there is $\ll 1$ electron per primitive unit cell) and when the band is almost full (i.e. when there are just less than two electrons per primitive unit cell).

Semiconductors and Insulators

6.1 Introduction

Thus far we have been mainly preoccupied with the properties of metals, and we shall return to them in later chapters. However, Section 3.3.2 indicated that certain substances with even numbers of valence electrons per unit cell may not conduct electricity at $T = 0$ because they possess the correct number of electrons to exactly fill a band, given that the next available empty states are somewhat removed in energy. Such materials are known as **semiconductors** or **insulators**; as they have proved to be technologically very important, we shall take some time to consider the general properties of their bandstructure. The concepts encountered will serve to illustrate the general principles which we have derived in the previous three chapters. For brevity I shall refer to such materials collectively as 'semiconductors'; as we shall see, the distinction between a semiconductor and an insulator is fairly nebulous.

The most important aspect of semiconductor bandstructure may be summarised as follows; at absolute zero the highest completely filled band (the **valence band**) is separated from the lowest empty band (the **conduction band**) by an energy gap or band gap E_g. Therefore the material does not conduct electricity at $T = 0$. At finite temperatures a variety of processes enable electrons to be excited into the conduction band and empty states to occur in the valence band, thus allowing electrical conduction. However, as will be shown in the rest of Chapter 6, the presence of the energy gap still dominates the properties of the semiconductor.

Most of the technologically-important semiconductors, such as Ge, Si, GaAs and (Hg,Cd)Te, have a face-centred cubic lattice with a two-atom basis; this has already been illustrated in Fig. 4.4. In the case of Si and Ge, the atoms on the A and B sites in Fig. 4.4 are identical (the so-called *diamond* structure). In the case of binary semiconductors such as (for example) GaAs the A sites will have Ga atoms and the B sites As atoms; the crystal is said to have the **zinc blende structure**. As stated above, in each case, the underlying lattice is face-centred cubic and there is a two atom basis. Thus, both the group IV elements and many of the binary semiconductors such as GaAs have the same shape of first Brillouin zone, which is shown in Fig. 4.4 and which will be useful for understanding the band diagrams in the following discussion.

6.1	Introduction	49
6.2	Bandstructure of Si and Ge	50
6.3	Bandstructure of the direct-gap III–V and II–VI semiconductors	53
6.4	Thermal population of bands in semiconductors	56

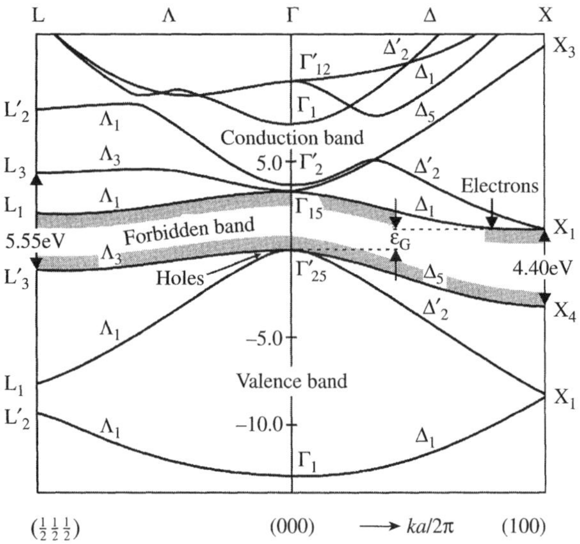

50 *Semiconductors and Insulators*

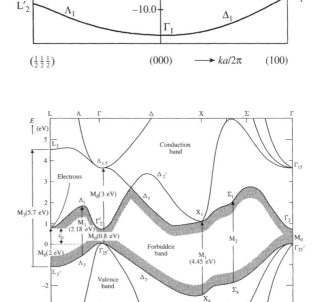

Fig. 6.1 Calculated bandstructure of Si projected along the (100) and (111) directions. The shading indicates the lowest (unoccupied at $T = 0$) conduction band and the highest (occupied at $T = 0$) valence band, with the 'forbidden band' or band gap in between. (Based on M. Cardona and F. Pollack, *Phys. Rev.* **142**, 530 (1966).)

Fig. 6.2 Calculation of the bandstructure of Ge along various projections. The shading indicates the lowest (unoccupied at $T = 0$) conduction band and the highest (occupied at $T = 0$) valence band. (Based on J.C. Philips et al., *Proc. int. conf. phys. semiconductors* (Exeter, UK) (The Institute of Physics, London, 1962) 564.)

6.2 Bandstructure of Si and Ge

6.2.1 General points

Si and Ge were the first semiconductors to be exploited commercially, and Si continues to be very important as a technological material.[1] Figures 6.1 and 6.2 show calculations of the bandstructures of Si and Ge respectively. At first sight, the situation looks a complex, fearful mess. However, at the typical temperatures that will concern us (0–$\sim$ 300 K), the only dramatic action (i.e. thermal excitation of electrons and holes, optical absorption edges) will occur close to the highest point in the valence band and the lowest points in the conduction band.

[1] See, for example, *Semiconductor physics*, by K. Seeger (Springer, Berlin, 1991) Section 5.7 and references therein.

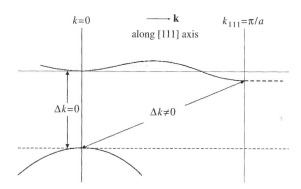

Fig. 6.3 Schematic of direct and indirect optical transitions in Ge.

6.2.2 Heavy and light holes

In both Si and Ge, two bands converge at the valence band maximum in the Brillouin-zone centre (the Γ-point). These bands are known as the **heavy-** and **light-hole bands**; the flatter one, with its large value of $(d^2E/dk^2)^{-1}$, is the heavy-hole band, and the steeper one the light-hole band (for obvious reasons—see Fig. 5.1).

The heavy holes tend to dominate the properties of the valence-band extremum; their heavier effective mass means that their density of states will be much larger than that of the light-holes (see Section 5.2.4 and Exercise 1.1).

6.2.3 Optical absorption

Photons with energy $h\nu \sim E_g \sim 0-2$ eV will have wavevectors which are $\ll$ typical Brillouin-zone size (energetic X-rays ($h\nu \gg 1$ eV) have wavevectors $\sim$ Brillouin-zone size). This means that transitions involving the absorption of a photon to excite an electron from the valence band to the conduction band, leaving a hole behind, will be essentially vertical in k-space. The transition strengths will be greatest when the joint density of initial and final states is large, i.e. when conduction and valence bands are approximately parallel.

Note that Si and Ge are **indirect-gap semiconductors**; the smallest band separation (the thermodynamic band gap, which determines the thermal population of electron and hole states) is not vertical in k-space. Figure 6.3 shows a schematic representation of this situation in Ge; strong optical transitions will occur at the point labelled $\Delta k = 0$ on the diagram, at a higher energy than the thermodynamic band gap, which is labelled $\Delta k \neq 0$.

It turns out that optical transitions *can* occur at energies close to the thermodynamic band gap in Si and Ge if both a phonon (see Appendix D) and a photon are involved; emission or (at higher temperatures) absorption of a phonon with the appropriate wavevector allows momentum to be conserved so that the transitions labelled $\Delta k \neq 0$ in Fig. 6.3 are possible.

Figure 6.4 shows the optical absorption coefficient of Ge at 300 K and 77 K. Both the indirect ($\Delta k \neq 0$) and direct ($\Delta k = 0$) transitions can be seen as absorption edges. The higher energy direct transition is some 100 times stronger than the indirect transition. In addition, the indirect transition is stronger at 300 K than at 77 K, as there is a substantial population of phonons of a suitable wavevector to take part in the transition at the higher

52 Semiconductors and Insulators

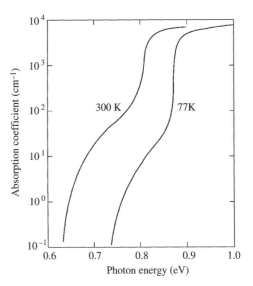

Fig. 6.4 Optical absorption of Ge at 300 K and 77 K. (Data from R. Newman and W.W. Tyler, *Solid state physics* **8**, 49 (1959).)

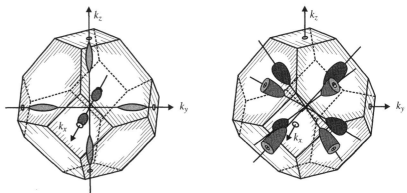

Fig. 6.5 Constant energy surfaces of Si (left) and Ge (right).

temperature; optical transitions assisted by phonon absorption thus becomes possible in addition to those assisted by emission.

Finally, note that the energy gaps change with temperature. This is not unexpected in models of bandstructure; e.g. in the tight-binding model of bands, thermal expansion will influence the atomic separation and hence the transfer integrals.

6.2.4 Constant energy surfaces in the conduction bands of Si and Ge

The conduction band minima in Si and Ge (see Figs. 6.1 and 6.2) are situated away from the Brillouin-zone centre (the Γ point) and are thus rather anisotropic. At energies close to (i.e. $<\sim$ 50 meV from) the bottom of the conduction bands, effective mass tensors with constant coefficients (as discussed in Exercise 5.2) can be used to describe the conduction-band minima in Si and Ge. By choosing suitable axes, the tensors can be made diagonal. The constant energy surfaces are approximate ellipsoids of rotation about one of these axes (see Fig. 6.5).

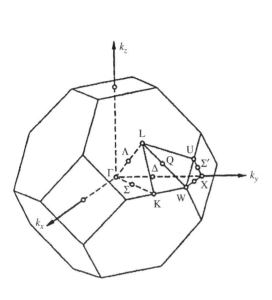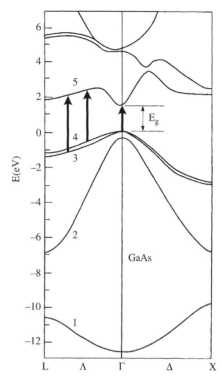

Fig. 6.6 Left-hand side: Brillouin zone of GaAs. Right-hand side: bandstructure of GaAs. The important bands are numbered; 2: spin-orbit split-off band, 3: light-hole band, 4: heavy-hole band, 5: conduction band. The arrows indicate the transitions referred to in Section 8.7.2. (After D. Long, *Energy bands in semiconductors* (Wiley, New York, 1968).)

6.3 Bandstructure of the direct-gap III–V and II–VI semiconductors

6.3.1 Introduction

Compound semiconductors such as GaAs and InSb are known as III–V semiconductors because they combine elements from groups III and V of the periodic table. In a similar way, semiconductors such as (Cd,Hg)Te are known as II–VI semiconductors. Many III–V and II–VI semiconductors have technological applications, often as emitters and detectors of electromagnetic radiation. It is important that we examine their bandstructure and consequent properties, in order to see why they are used for such applications. We shall see that one of the most important points is that the minimum energy band gaps in most of these materials are direct, so that optical transitions with $\Delta k \approx 0$ can occur.

6.3.2 General points

Figure 6.6 shows the Brillouin zone and calculated bandstructure of GaAs, a typical direct-gap III–V semiconductor. Note that

- both heavy- and light-hole bands are present at the valence-band extremum, as before (cf. Figure 6.2);

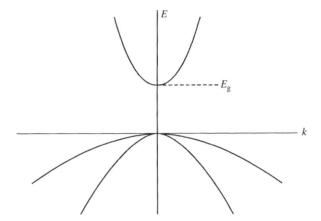

Fig. 6.7 Schematic of bandstructure of GaAs close to the direct band-gap. The origin of the energy scale is taken as the top of the valence band.

- the thermodynamic band gap separating the highest point of the valence band from the lowest point of the conduction band is direct;
- Just below the valence-band maximum, there is another band, known as the spin-orbit split-off band.

For most practical purposes, we need consider only the electronic dispersion relationships close to the valence band maximum and conduction band minimum; a schematic of this region of k-space is given in Fig. 6.7. In fact, Fig. 6.7 is a good representation of this region for many direct-gap III–V (e.g. InSb, InAs) and II–VI (e.g. CdTe, $Cd_{0.2}Hg_{0.8}Te$) semiconductors. Table 6.1 shows band edge effective masses for three examples. Note that

- the light-hole effective mass $m^*_{lh} \approx m^*_c$, the electron (conduction-band) effective mass;
- m^*_c and m^*_{lh} scale roughly with E_g;
- the heavy-hole effective mass m^*_{hh} is almost material-independent.

Table 6.1 Energy gaps (in eV), and effective masses (in units m_e) of typical III–V semiconductors.

Material	E_g	m^*_c	m^*_{lh}	m^*_{hh}
GaAs	1.52	0.067	0.082	0.45
InAs	0.42	0.023	0.025	0.41
InSb	0.24	0.014	0.016	0.40

The fact that wider band gap semiconductors have larger conduction-band and light-hole effective masses is rather a general feature of semiconductors and insulators; it will be an important factor in the discussion below.

6.3.3 Optical absorption and excitons

Figure 6.8 shows the optical absorption coefficient of GaAs at several temperatures. Note that the lowest energy band gap is direct ($\Delta k = 0$); in contrast to the case of Ge, there is no indirect band gap absorption at energies below the direct gap. There is, nevertheless, one marked similarity to the case of Ge; the direct energy-gap rises as the temperature falls.

At lower temperatures, an extra peak emerges close to the onset of absorption. This is due to the formation of an **exciton**, a bound electron-heavy-hole pair.[2] (As stated above, the heavy holes dominate the spectra because of their large density of states.) We shall look at some of their general properties in the following section.

[2] Excitons are also seen just below the direct band gaps of semiconductors such as Ge and Si at sufficiently low temperatures; see e.g. T.P. McLean, in *Progress in semiconductors*, vol. 5 (Wiley, 1960).

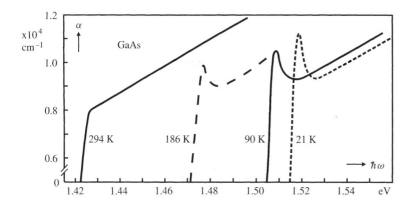

Fig. 6.8 Optical absorption coefficient of GaAs at several different temperatures. Note that the joint density of states of the electrons and holes might be expected to give an absorption coefficient which is initially proportional to $(\hbar\omega - E_g)^{1/2}$ (see Section 5.2.4), as the conduction and valence bands are almost parabolic close to the band-edges. However, this is masked by the contribution of the exciton to the joint density of states. (Data from M.D. Sturge, *Phys. Rev.* **127**, 768 (1962).)

6.3.4 Excitons

At $T \leq 200$ K there will be very few carriers around in the conduction and valence bands of GaAs and therefore little screening. An electron and hole created by the absorption of a photon will therefore interact via the Coulomb force. The situation is analogous to the hydrogen atom (see Appendix E), but with the following adjustments.

- The mass of the electron in the hydrogen atom is just the free-electron mass m_e; furthermore, the proton (i.e. hydrogen nucleus) is much heavier than the electron, so that its motion is almost negligible. However, in an exciton, the electron and heavy-hole possess **effective masses** which are often within a factor ten of each other; both take part in the 'orbital' motion. Therefore m_e in the hydrogen calculation must be replaced by the **reduced effective mass** μ^*, with

$$\frac{1}{\mu^*} = \frac{1}{m_c^*} + \frac{1}{m_{hh}^*} \tag{6.1}$$

- The electron in hydrogen moves *in vacuo*. In contrast, the electron and hole in the exciton move within a medium with relative permittivity ϵ_r.

Thus the energy levels of the exciton are given by

$$E(n) = \frac{\mu^*}{m_e} \frac{1}{\epsilon_r^2} \times \frac{13.6 \text{ eV}}{n^2}, \tag{6.2}$$

with n an integer.[3] Here $n = 1$ represents the ground state (the 1s state) of the exciton, and $n = \infty$ represents a free electron–hole pair (the 'ionisation' of the exciton). Substituting typical values for ϵ_r ($\epsilon_r \sim 10 - 15$ for many semiconductors) and μ^* (e.g. $\mu^* \approx 0.06 m_e$ in GaAs) gives an exciton binding energy $E(1) \approx 5$ meV. A similar approach yields an effective Bohr radius for the exciton of $>\sim 200$ nm (i.e. $>\sim 60$ atomic spacings), thus justifying the approach followed for working out the excitonic energy levels (i.e. that the hole and electron move through a macroscopic 'medium' described by properties such as ϵ_r and μ^*). The GaAs spectra shown above (see Fig. 6.8) are dominated by the 1s ($n = 1$) state of the exciton. However, in materials with wide band-gaps (i.e. large μ^*; see Section 6.3.2) and low values of ϵ_r,

[3] I have used the conventional symbol for the quantum number n so as not to confuse any atomic physicists reading this book. This is the same as the standard symbol n for electron density, again used throughout this book so as not to confuse semiconductor physicists. The context for each n is plainly stated; they do not occur in the same equation and so I hope that it is obvious which is which.

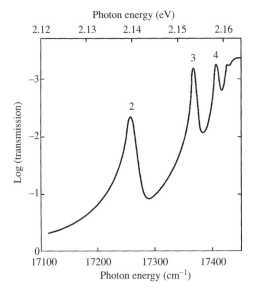

Fig. 6.9 Absorption of the wide-gap semiconductor Cu_2O, showing hydrogen-like states of the exciton; the quantum number n of each state is labelled. (Data from S. Nikitine *et al.*, *Proc. int. conf. phys. semiconductors* (Exeter, UK) (The Institute of Physics, London, 1962).)

the exciton binding energies are much greater, and the individual hydrogen-like states of the exciton can be clearly seen in absorption data; an example is shown in Fig. 6.9. Conversely, in narrow-gap semiconductors with small effective masses, such as InSb (see Section 6.3.2), the exciton binding energies are small and excitonic effects are rarely seen.

6.3.5 Constant energy surfaces in direct-gap III–V semiconductors

Figure 6.6 shows that the conduction-band minimum and valence-band maximum of GaAs are at the Brillouin-zone centre (the Γ point), a point of very high symmetry. Close to these points, the bands are quite isotropic (especially the conduction band); for many purposes, each band can therefore be treated using a single, constant effective mass (see Table 6.1) to a reasonable degree of accuracy, as long as the carrier energies do not exceed a few tens of meV.

6.4 Thermal population of bands in semiconductors

6.4.1 The law of mass action

The most notable thing about semiconductors (compared to metals) is the fact that the number of mobile charge carriers varies with temperature. We shall now derive the equation governing this behaviour. The number density of electrons dn with energy between E and $E + dE$ will be given by

$$dn = f_D(E, T)g(E)dE, \tag{6.3}$$

where f_D is the Fermi–Dirac distribution function

$$f_D = \frac{1}{e^{\frac{(E-\mu)}{k_BT}} + 1} \tag{6.4}$$

$g(E)$ is the density of states and μ is the chemical potential.[4]

Distribution functions for electrons and holes Now let us consider the case that μ is within the band gap, such that the chance of occupancy of a conduction-band state is $\ll 1$. This implies that $(E - \mu) \gg k_BT$, so that the Fermi–Dirac distribution function f_{DC} for an electron in the conduction band can be approximated as

$$f_{DC} \approx \frac{1}{e^{\frac{(E-\mu)}{k_BT}}} = e^{-\frac{(E-\mu)}{k_BT}}. \tag{6.5}$$

[4] In the next bit of the derivation we are going make a distinction between the Fermi–Dirac distribution function in the valence and conduction bands, as though it becomes two different functions. Of course it does not; however, differing approximations can be made for the two bands, so that some sort of short-hand labelling which distinguishes the two is desirable.

The chance of finding a hole in the valence band will be $1 - f_{DV}$, where f_{DV} is the Fermi–Dirac distribution function within the valence band. Under similar conditions to those used in the derivation of eqn 6.5, $(E - \mu)$ will be negative and $|E - \mu| \gg k_BT$. Therefore the approximation

$$1 - f_{DV} = 1 - (1 + e^{\frac{(E-\mu)}{k_BT}})^{-1} \approx 1 - (1 - e^{\frac{(E-\mu)}{k_BT}}) = e^{-\frac{(\mu-E)}{k_BT}} \tag{6.6}$$

can be made.

Density of states We assume that the conduction and valence bands near the band-edges are parabolic, each characterised by a single effective mass. (Even though there are both heavy and light holes, this is not a bad approximation, as the heavy holes have a much bigger effective mass and hence density of states; the light holes can be neglected.) The densities of states for the conduction band (g_c) and valence band (g_v) are therefore

$$g_c = C m_c^{*\frac{3}{2}} (E - E_c)^{\frac{1}{2}} \tag{6.7}$$

and

$$g_v = C m_{hh}^{*\frac{3}{2}} (E_v - E)^{\frac{1}{2}} \tag{6.8}$$

respectively, where m_c^* and m_{hh}^* are the electron and heavy-hole effective masses respectively, E_c and E_v are the energies of the conduction and valence band edges respectively and $C = (1/2\pi^2)(2/\hbar^2)^{3/2}$ (see Section 5.2.4).

Combining eqns 6.3, 6.5 and 6.7, the number of electrons in the conduction band per unit volume is

$$n = \int_{E_c}^{\infty} f_{DC} g_c dE \approx C m_c^{*\frac{3}{2}} \int_{E_c}^{\infty} (E - E_c)^{\frac{1}{2}} e^{-\frac{(E-\mu)}{k_BT}} dE. \tag{6.9}$$

Making the substitution $y = (E - E_c)/(k_BT)$ gives

$$n \approx C(m_c^* k_B T)^{\frac{3}{2}} e^{-\frac{(E_c-\mu)}{k_BT}} \int_0^{\infty} y^{\frac{1}{2}} e^{-y} dy, \tag{6.10}$$

leading to

$$n = 2\frac{(2\pi m_c^* k_B T)^{\frac{3}{2}}}{h^3} e^{-\frac{(E_c-\mu)}{k_BT}} = N_c e^{-\frac{(E_c-\mu)}{k_BT}}. \tag{6.11}$$

Here N_c is the effective number density of *accessible* states at the conduction band bottom.

Similarly, combining eqns 6.3, 6.6 and 6.8, p, the number of holes in the valence band per unit volume is

$$p = \int_{-\infty}^{E_v} (1 - f_{DV}) g_v dE \approx C m_{hh}^{*\frac{3}{2}} \int_{-\infty}^{E_v} (E_v - E)^{\frac{1}{2}} e^{-\frac{(\mu - E)}{k_B T}} dE. \quad (6.12)$$

The same type of substitution as in the case of the electrons leads to

$$p \approx 2 \frac{(2\pi m_{hh}^* k_B T)^{\frac{3}{2}}}{h^3} e^{-\frac{(\mu - E_v)}{k_B T}} = N_v e^{-\frac{(\mu - E_v)}{k_B T}}, \quad (6.13)$$

where N_v is the effective number density of *accessible* states at the valence band top.

Combining eqns 6.11 and 6.13 gives

$$np \approx N_c N_v e^{-\frac{E_g}{k_B T}} = W T^3 e^{-\frac{E_g}{k_B T}}, \quad (6.14)$$

where the energy gap $E_g = E_c - E_v$, and W is a constant depending on the details of the valence and conduction-band extrema. This is known as the **law of mass action**. The exponential dependence of np on the temperature is the chief mechanism determining the temperature dependence of the electrical conductivity of semiconductors (see Chapter 9).

6.4.2 The motion of the chemical potential

We now consider what happens to the chemical potential as a function of T. Setting $n = p$ and equating eqns 6.11 and 6.13 gives

$$\frac{N_v}{N_c} = e^{\frac{(2\mu - E_c - E_v)}{k_B T}}. \quad (6.15)$$

Now $(N_c/N_v) = (m_c^*/m_{hh}^*)^{3/2}$ (see Section 5.2.4), so that the chemical potential is given by

$$\mu = \frac{1}{2}(E_c + E_v) + \frac{3}{4} k_B T \ln \left(\frac{m_{hh}^*}{m_c^*} \right). \quad (6.16)$$

Therefore μ starts off in the middle of the band-gap (i.e. exactly half way between the empty and full states) at $T = 0$; as m_{hh}^* is bigger than m_c^* in most semiconductors, μ usually moves up as T rises.

6.4.3 Intrinsic carrier density

Let us first assume that the only source of electrons in the conduction band and holes in the valence band is the thermal excitation of electrons across the band gap (i.e. there are no impurities; see Section 6.4.4 below); such electrons and holes are known as *intrinsic* carriers. In this case (see eqn 6.14)

$$n_i = n = p = T^{\frac{3}{2}} W^{\frac{1}{2}} e^{-\frac{E_g}{2 k_B T}}, \quad (6.17)$$

6.4 Thermal population of bands in semiconductors

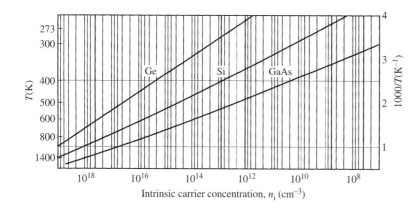

Fig. 6.10 Calculated intrinsic carrier densities n_i versus temperature in Ge ($E_g = 0.74$ eV), Si ($E_g = 1.17$ eV) and GaAs ($E_g = 1.52$ eV).

where n_i is the **intrinsic carrier density**. Note that the exponential contains $E_g/2$ and *not* E_g; this is because the creation of each electron in the conduction band automatically makes a hole in the valence band.

Figure 6.10 shows n_i for Ge, Si and GaAs as a function of T; the exponential dependence on E_g means that relatively small differences in band gap result in n_is which are several orders of magnitude apart.

6.4.4 Impurities and extrinsic carriers

For virtually all practical applications (i.e. those at room temperature), the conductivity of semiconductors such as GaAs, Ge and Si is dominated by **extrinsic carriers**, those provided by doping the semiconductor with impurities.

Two types of impurity will concern us, **donors**, which donate electrons, and **acceptors**, which provide holes.

- **Donors** An atom of the semiconductor with valence $\mathcal{V}$ is replaced by a donor atom, with valence $\mathcal{V} + 1$. After the donor has bonded into the host lattice, it will still have one electron left over. Examples of donors include P in Si or Ge, Si on a Ga site in GaAs, and Te on an As site in GaAs.
- **Acceptors** An atom of the semiconductor with valence $\mathcal{V}$ is replaced by an acceptor atom, with valence $\mathcal{V} - 1$. Bonding into the lattice will therefore leave the acceptor short of one electron, i.e. with a surplus hole. Typical acceptors include Al and Ga in Si and Ge, Be on a Ga site in GaAs and Si on an As site in GaAs.

The surplus carrier (electron or hole) will orbit the (positively or negatively) charged ionic core of the impurity; again (see Section 6.3.4) the situation is analogous to hydrogen, with the following alterations (cf. Section 6.3.4)

- the free-electron mass m_e in the hydrogen is replaced by the effective mass of the carrier;
- the carrier moves through a medium of relative permittivity ϵ_r.

Thus, the energy levels of e.g. a donor are given by

$$E(n) = \frac{m_c^*}{m_e} \frac{1}{\epsilon_r^2} \times \frac{13.6 \text{ eV}}{n^2}, \qquad (6.18)$$

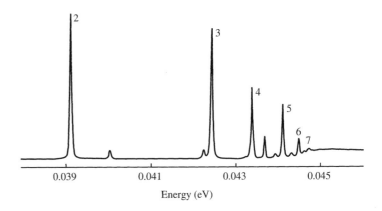

Fig. 6.11 Far-infrared absorption due to P impurities (donors) in Si. Hydrogen-like transitions between the $n = 1$ ground-state and higher levels can clearly be seen. (Data courtesy of Prof. R.A. Stradling.)

[5] The perceptive reader will notice that the transitions are *not quite* at the energies predicted by the simple model of hydrogen. In Ge and Si, a further complication arises from the multi-valley nature of the conduction bands; in Si, for example, there are six impurity states from six minima. These interact so that the final pattern of levels observed is not precisely hydrogenic (the situation in direct-gap semiconductors with a single, zone-centre conduction-band minimum (e.g. GaAs) is rather closer to the hydrogenic model). Similarly, the degeneracy of the valence band modifies the simple hydrogenic acceptor levels.

with n an integer (see Fig. 6.11 for a graphic illustration of these energy levels).[5]

As in the exciton model, $n = 1$ represents the ground state (the 1s state) of the donor, and $n = \infty$ represents a free electron in the conduction band (the 'ionisation' of the donor). Similar considerations hold for acceptors.

Substituting typical values for ϵ_r ($\epsilon_r \sim 10 - 15$ for many semiconductors) and m_c^* (e.g. $m_c^* \approx 0.07 m_e$ in GaAs) into eqn 6.18 gives an impurity binding energy $E(1) \approx 5$ meV. Semiconductors with wide band gaps will tend to have electrons and light holes with heavier effective masses, resulting in larger impurity binding energies, whilst narrow-gap semiconductors will tend to have small impurity binding energies (see Section 6.3.2).

The hydrogen-like approach also yields an effective Bohr radius $a_B^* \sim 10$ nm. The large Bohr radius of the donor implies that the donated electron will spend very little time close to its 'parent' ionic core. Therefore the impurity binding energy of a donor will be almost entirely determined by the host semiconductor rather than the parent ion.

6.4.5 Extrinsic carrier density

As long as the numbers of donors and/or acceptors are small enough such that the chemical potential remains in the band-gap (i.e. the chance of occupancy of an electron state in the conduction band or a hole state in the valence band is much less than 1), the derivation of Section 6.4.1 will be applicable, so that the law of mass action will still hold.

In order to find n and p when impurities are present we therefore use eqn 6.14

$$np = T^3 W e^{-\frac{E_g}{k_B T}}$$

combined with the conservation law

$$n - p = N_D - N_A, \qquad (6.19)$$

where N_D is the density of donors and N_A is the density of acceptors (both are assumed to only provide one carrier each). The right-hand side of eqn 6.19 shows that the presence of donors can be **compensated** by acceptors; this is known (surprise, surprise!) as **compensation**.

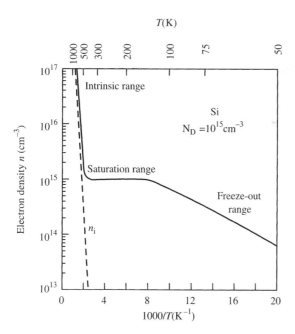

Fig. 6.12 Temperature dependence of the electron density in Si with a net donor density of $N_D - N_A = 10^{15} \text{cm}^{-3}$.

Binding energies of order 5–10 meV imply that most impurities in semiconductors such as GaAs and Si will be ionised at room temperature.[6] Figure 6.12 shows the temperature dependence of the electron density in the conduction band of Si with a net donor density of $N_D - N_A = 10^{15} \text{cm}^{-3}$. At cryogenic temperatures the extrinsic electrons *freeze out* onto the donors; however above about 200 K, all of the donors are ionised (the saturation range) and n is relatively constant. At ~ 500 K the intrinsic contribution to n becomes non-negligible and then rapidly dominates.

A comparison of Figs 6.10 and 6.12 shows that the n_i in GaAs and Si are very low indeed at room temperature, much less than practically-achievable values of $N_D - N_A$; samples of these semiconductors would have to be free of donors to better than about 1 part in $10^{12} - 10^{15}$ (impossible!) for one to see something approaching true intrinsic behaviour.

Finally, we examine the behaviour of the chemical potential in the presence of impurities. To see where μ goes, we repeat the derivation of Section 6.4.2, but using the number density of empty donor states ($N_D e^{-(\mu - E_D)/k_B T}$), where E_D is the energy of the donor ground-state) in place of the number of holes in the valence band.[7] (We are thus considering an n-type semiconductor; however, an equivalent derivation for a p-type semiconductor uses the same ideas.) This gives

$$\mu = \frac{1}{2}(E_D + E_c) + \frac{1}{2}k_B T \ln\left(\frac{N_D}{N_c}\right), \qquad (6.20)$$

i.e. μ can either rise or fall with increasing temperature.[8]

[6] For moderate values of N_D (see Section 6.4.6), this implies that $(E_D - \mu) \gg k_B T$ and $(\mu - E_A) \gg k_B T$, where E_D and E_A are the energies of the donor and acceptor ground-states respectively. Therefore μ will not only be far from the band-edges, it must also be far from the impurity levels.

[7] So as not to complicate the discussion, the expression used for the number of empty donor states is a simplification. Those intending to embark on calculations or research should note that the number should be $N_D(1 - 1/(1 + g^{-1}e^{(\mu - E_D)/k_B T}))$, where g is the degeneracy of the donor ground-state (cf. eqn C.22). To see where g comes from, consider the 1s state of a hydrogenic donor. This is technically doubly degenerate owing to the fact that electrons can have spin up or spin down. Double occupancy of the donor is not energetically favourable; however, the statistical mechanics 'knows' that there are two states available at that energy. See, for example, *Semiconductor physics*, by K. Seeger (Springer, Berlin, 1991).

[8] Beware! The derivation of this equation assumes that $E_c - \mu$ and $\mu - E_D$ are both much greater than $k_B T$. We have already seen that this is unlikely to be true at room temperature; therefore this formula is only valid at much lower temperatures. It is, nevertheless, a useful guide to the low-temperature movement of μ.

6.4.6 Degenerate semiconductors

Once all of the donors are ionised (the exhaustion regime), $n = N_D$; equating eqn 6.11 to N_D gives

$$\mu \approx E_c + k_B T \ln(\frac{N_D}{N_c}), \qquad (6.21)$$

where the $\approx$ re-emphasises that we are approaching a dodgy situtation as far as the approximations used to derive eqn 6.5 are concerned. There are two cases to note.

- $N_D < N_c$ leads to a chemical potential which is still in the band-gap. The number of electrons in the conduction band is still small compared to the number of available states, so that the quasiclassical statistics used in eqn 6.5 and the derivation that follows it are probably valid. In this case, the semiconductor is **non-degenerate**.
- $N_D > N_c$ leads to a chemical potential in the conduction band. Therefore we cannot use the quasiclassical approximation for the Fermi–Dirac distribution function. The semiconductor behaves somewhat like a metal and is said to be **degenerate**.

6.4.7 Impurity bands

If a large enough concentration of impurities is placed in a semiconductor, the wavefunctions of the carriers bound to the impurities will start to overlap; the criterion for this for *e.g* donors is $N_D \geq \sim 1/a_B^{*3}$. The situation becomes rather similar to the tight-binding model, and carriers will be able to move from impurity to impurity. The semiconductor will remain electrically conductive even at very low temperatures; it is another instance of a **degenerate semiconductor**. However, in contrast to the case discussed in the previous section, the chemical potential will depend rather less on the temperature; we will have a situation much more analogous to a (rather dirty) conventional metal.

6.4.8 Is it a semiconductor or an insulator?

Most textbooks state that insulators are just wide-gap semiconductors, the implication being that thermal effects excite a negligible number of electron–hole pairs across the band-gap if it is wide (see eqn 6.14).

However, this picture is an over-simplification; as we have seen, the carriers which populate the bands at room temperature come mostly from impurities. Section 6.3.2 stated that carrier effective masses become heavier as the band-gap increases; as the effective mass increases, the impurity binding energy will increase (see eqn 6.18). As the gap widens, eventually the impurity will become so deep that it will not be ionised at room temperature.

The distinction between wide-gap semiconductors and insulators is very blurred; e.g. studies of doped diamond for use in potential ultraviolet/blue LEDs are carried out, the aim being to find donors and acceptors which are shallow enough not to lead to carrier freeze-out at sensible operating temperatures of $\sim$ 300 K. Conversely, much work has been done to dope semiconductors like GaAs with so-called **deep** traps, impurity states which are deliberately made so deep that the electrons and holes will remain frozen on them even at room temperature. Excess impurities will 'soak up' any other carriers that are around, giving very high resistivities indeed. Such **semi-insulating** material is often used as a substrate to support devices made from more highly-conductive GaAs and (Ga,Al)As (see Section 7.3).

6.4.9 A note on photoconductivity

It has been mentioned that semiconductors are often employed to detect photons. The detection involves **photoconductivity**, which can occur via three mechanisms;

(1) the creation of electron–hole pairs through interband (intrinsic) absorption of radiation; typically this occurs at photon energies $\sim E_g \sim 1$ eV or wavelengths $\lambda \sim 1$ μm, although Chapter 7 will show that by alloying semiconductors such as CdTe with semimetals such as HgTe, materials with band gaps as small as 0.1 eV (corresponding to $\lambda \sim 10$ μm) can be made;
(2) the excitation of electrons (or holes) from donor (acceptor) impurities into the conduction (valence) band; typically this occurs at energies $\sim$ 5 meV to 50 meV, corresponding to $\lambda \sim 200 - 20$ μm;
(3) free-carrier absorption.

In cases 1 and 2, the carrier numbers are increased; in case 3, the number of carriers remains constant but the conductivity changes because of a change in mobility. The latter mechanism is usually important in cases where the carrier effective mass is low and the band is strongly nonparabolic[9] as the low density of band states (from the small effective mass) results in long relaxation times for the excited carriers. Free-carrier absorption is proportional to the real part of the conductivity $\sigma = \sigma_0/(1 + i\omega\tau)$, and so increases rapidly at low energies/long wavelengths.

[9]This is the case in the conduction bands of narrow-gap semiconductors such as $Cd_{0.2}Hg_{0.8}Te$ and InSb.

Further reading

Useful alternative treatments are given in *Electricity and magnetism*, by B.I. Bleaney and B. Bleaney, third edition (Oxford University Press, Oxford, 1989), Chapters 17–19, *Solid state physics*, by G. Burns (Academic Press, Boston, 1995), Sections 10.10–10.20 and Chapter 18, and *Low-dimensional semiconductor structures*, by M.J. Kelly (Clarendon Press, Oxford, 1995), Chapter 1 and Sections 2.1–2.3, 2.7, 4.4, 5.1–5.3, 9.1–9.2. Optical properties of semiconductors are tackled in depth in *Optical properties of solids*, by M. Fox (OUP 2001).

Essentials of semiconductor physics by Tom Wenckebach (Wiley, New York, 1999) provides a slightly offbeat, but algebraically rigorous treatment of many of the properties of semiconductors which would be useful for those embarking on research or calculations. A similarly detailed but more traditional approach is given in *Semiconductor physics*, by K. Seeger (Springer, Berlin, 1991)and *Fundamentals of semiconductors*, by P. Yu and M. Cardona (Springer, Berlin, 1996). The methods used to calculate the bandstructure of semiconductors have been reviewed by F. Bassani (Chapter 2 of *Semiconductors and semimetals*, Volume 1, edited by R.K. Willardson and A.C. Beer

(Academic Press, New York, 1966)) and E.O. Kane (Chapter 4 of *Handbook on semiconductors*, volume 1, edited by T.S. Moss (North Holand, Amsterdam, 1982)); the latter reference includes the famous **k.p** method for evaluating the bandstructure close to band extrema (see also *Fundamentals of semiconductors*, by P. Yu and M. Cardona (Springer, Berlin, 1996), Chapter 2.).

A very wide range of properties of semiconductors (band gaps, effective masses etc.) can be looked up in the comprehensive Landholt-Börnstein series, *Numerical data and functional relationships in science and technology*, New Series, Group III, Volume 17, edited by O. Madelung (Springer, Berlin, 1982).

Impurities in semiconductors are dealt with comprehensively in the mammoth tome *Doping in III–V semiconductors*, by E.F. Schubert (Cambridge University Press, 1993).

Semiconductor devices are reviewed in *Modern semiconductor device physics* (1997), *Physics of semiconductor devices* (second edition, 1981) and *Semiconductor devices: physics and technology* (1985), all by S.M. Sze (Wiley, New York).

Exercises

(Note that there are many more problems about the properties of semiconductors in Chapters 7, 10 11 and 12.)

(6.1) An intrinsic sample of GaAs is at a temperature of 300 K. Estimate the change in temperature required to increase its electrical conductivity by 10 per cent.

(6.2) Estimate the intrinsic carrier density in Ge at room temperature. Hence estimate the maximum impurity concentration allowable if Ge is to behave as an intrinsic semiconductor at room temperature. Are such levels of purity feasible?

(6.3) Later in the book, we shall see that the electrical conductivity σ of a semiconductor is given by

$$\sigma = ne\mu_c + pe\mu_{hh}$$

(eqn 9.11), where μ_c and μ_{hh} are the electron and hole mobilities respectively. Show that in a semiconductor with $\mu_c = \mu_{hh}$ the minimum electrical conductivity occurs for **intrinsic** material. In what applications might it be useful to have minimum conductivity?

(6.4) A sample of silicon has one part in 10^9 of its silicon atoms replaced by donor atoms. Find the position of the chemical potential μ at room temperature relative to the valence-band-edge and compare the carrier density with n_i, the intrinsic carrier density.

(6.5)* Imagine that the conduction band states of a semiconductor are completely full up to an energy xk_BT above the band edge and completely empty above this. Adjust the constant x to make the carrier density $n = N_c$, where N_c is defined in eqn 6.11. Hence, check the correctness of the description of N_c as the effective number density of accessible states.

(6.6) The ionisation energies E_D of the donors As, Sb and P in germanium are 14.0 meV, 9.8 meV and 12.8 meV respectively. Account for the order of magnitude of E_D and for the fact that it is not strongly dependent on the particular donor impurity.

How would you demonstrate experimentally that a donor which is not ionised has a range of excited energy levels?

Bandstructure engineering

7.1 Introduction

The idea of making artificial solids with bandstructures tailored to particular applications has long fascinated condensed matter physicists. In this chapter, we shall explore some of the ways in which bands with desired properties can be engineered using what has been termed 'chemical architecture'. A very simple example of this is the use of **semiconductor alloys**, in which a wide-gap semiconductor and a narrow-gap semiconductor are combined to give a substance with a desired intermediate band-gap. A second example is the **semiconductor superlattice** or **heterostructure**; here, very thin layers of different semiconductors are superimposed. A third approach is to use small organic or inorganic molecular units to build up solids with desirable metallic, semiconducting or even superconducting properties. We shall treat each of these cases in turn, starting with the semiconductor systems.

7.1	Introduction	65
7.2	Semiconductor alloys	65
7.3	Artificial structures	66
7.4	Band engineering using organic molecules	75
7.5	Layered conducting oxides	78
7.6	The Peierls transition	81

7.2 Semiconductor alloys

The previous chapter mentioned the fact that direct-gap semiconductors are used in opto-electronic applications. Many of these applications involve the emission of light by recombination of an electron and a hole across the band-gap, or absorption of light of the band-gap energy to create an electron–hole pair and hence produce some form of photoconduction;[1] The size of the band-gap is therefore of great importance.

Figure 7.1 shows the band-gaps and lattice parameters for some of the more common elemental and binary semiconductors. The band-gaps available are not generally optimised for practical devices. A desire to improve on this has led to the field of **bandstructure engineering**, where a variety of artificial structures are used to provide bandstructure optimised to a particular application.

The simplest bandstructure engineering involves making **ternary alloys** such as (Ga,Al)As and (Hg,Cd)Te in order to achieve a desired band gap. I list three examples of the uses of such alloys.

$Ga_{1-x}Al_xAs$ is technologically important because GaAs and AlAs form a solid solution over the entire ($0 \leq x \leq 1$) composition range with very little variation of lattice parameter (see Fig. 7.1); this means that multilayers of very high quality can be fabricated (see below). A variable direct band-gap is obtained for the range $0 \leq x \leq\sim 0.35$; for $x \geq\sim 0.35$ the band-gap is indirect.

[1] See e.g. *Semiconductor physics*, by K. Seeger (Springer, Berlin, 1991), Chapters 12 and 13; *Optical properties of solids* by M. Fox, (Oxford University Press, Oxford, 2001), *Low-dimensional semiconductor structures*, by M.J. Kelly (Clarendon Press, Oxford, 1995), Chapter 18.

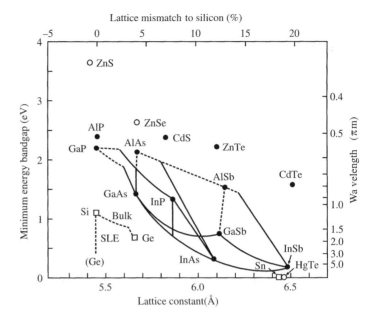

Fig. 7.1 Band-gap versus lattice parameter for some of the more common semiconductors. The curves indicate commonly-used alloys such as (Ga,Al)As, (Ga,In)As etc.; solid curves represent direct gaps and dashed curves indirect gaps.

The band-gap of $Ga_{1-x}In_xAs$ may be adjusted over an energy range which coincides with the low-attenuation region of many optical fibres.

$Hg_{1-x}Cd_xTe$ continues to be of great importance in the fabrication of infrared detectors covering the 10 μm ($x = 0.2$; $E_g \approx 100$ meV) and 5 μm ($x = 0.3$; $E_g \approx 200$ meV) atmospheric windows (regions where the atmosphere has little absorption). The 10 μm region also contains the peak thermal emission of ~ 300 K things, including human beings, so that there is a variety of medical (thermal imaging), meteorological and more sinister applications. Exhausts and jet engines emit well in the 5 μm range, and so the applications of $Hg_{1-x}Cd_xTe$ with $x = 0.3$ can be easily imagined.

7.3 Artificial structures

We now turn to a variety of structures which can be grown using techniques developed over the past twenty years or so.

7.3.1 Growth of semiconductor multilayers

The growth of high-quality semiconductor multilayers is described as **epitaxial** (the word is derived from the Greek words *epi* (upon) and *taxis* (arrangement)). The implication is that layers grow on a suitable single-crystal substrate, continuing the crystal structure of that substrate. The layers are thus supposed to be crystallographically well ordered. Two techniques are commonly used to grow epitaxial layers of semiconductors, **Molecular beam epitaxy** (MBE) and **Metal-organic vapour-phase epitaxy** (MOVPE; also known as OMVPE, MOCVD and OMCVD).

- **MBE** Figure 7.2 shows the main components of an MBE machine. The elements which make up the semiconductors to be grown evaporate from

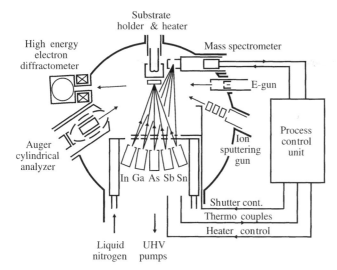

Fig. 7.2 Schematic of an MBE machine.

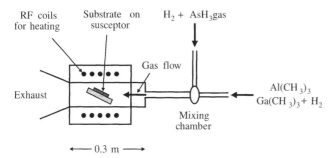

Fig. 7.3 Schematic of an MOVPE machine.

Knudsen cells at a rate controlled by the cell temperature; cells are also provided for dopants. The evaporated atoms form a beam which travels towards the substrate. Typically the evaporation rates are such that about one monolayer is deposited per second. The composition of the layers is controlled by shutters which can swing in front of each cell in about 50 ms, cutting off the beam from that cell. The substrate is kept at a well-defined temperature to ensure that the deposited atoms are reasonably mobile, so that they spread out over the substrate in monolayers rather than forming clusters. The whole chamber must be very well evacuated to prevent spurious dopants entering the layers; typical pressures in an MBE machine are $\sim 10^{-11}$ mbar. A number of *in situ* diagnostic techniques to monitor the growth of the multilayer are provided.

- **MOVPE** Figure 7.3 shows a schematic of an MOVPE machine. The components of the semiconductor to be grown travel to the substrate as gaseous precursors formed by reacting the elements with organic radicals; the precursors are carried along in a stream of hydrogen. Close to the substrate, which is heated by a radiofrequency coil, the precursors react, depositing the semiconductor on the substrate. A typical reaction is

$$\mathrm{Ga(CH_3)_3 + AsH_3 \rightarrow 3CH_4 + GaAs.} \qquad (7.1)$$

68 Bandstructure engineering

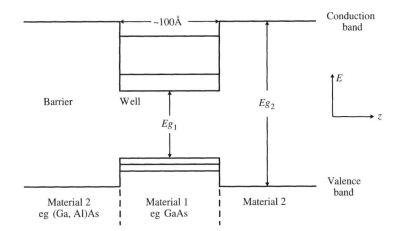

Fig. 7.4 Band-edges in a quantum well as a function of distance in the growth direction z. The energies of the subbands within the well are shown schematically.

Very fast solenoid valves enable the sources of the various components to be switched on and off rapidly.

7.3.2 Substrate and buffer layer

The substrate will be a high quality single crystal, usually of one of the materials to be included in the multilayer. Often the substrate will be **semi-insulating** GaAs (see Section 6.4.8). The first layer to be grown is called the *buffer layer*; it is often the same material as the substrate (but undoped), and is designed to 'smooth out' the lumps and bumps of the latter, to provide an atomically-smooth top surface on which to grow the active parts of the multilayer. More recently, **superlattice buffers** consisting of alternating thin (~ 1–2 nm) layers of e.g. GaAs and AlAs have sometimes been used for this purpose.

7.3.3 Quantum wells

The simplest multilayer system or **heterostructure** (*hetero* = more than one; i.e. the structure is made up of more than one semiconductor) is the **quantum well**. A thick (several hundred nanometres) layer of a wider gap material such as (Ga,Al)As is grown (the barrier material), followed by a $\sim 2 - 100$ nm layer of a narrower gap material such as GaAs (the well), followed by another thick layer of wider gap material. Figure 7.4 shows the resulting conduction and valence-band-edges; the relative heights of the discontinuities in the conduction- and valence-band-edges are instrinsic properties of the two materials involved.

The narrower-gap material forms a one-dimensional potential well in the conduction and valence bands; thus, in the z (growth) direction, the electron and hole levels are bound states of the well, known as **subbands**. The well will contain three sets of subbands,

- the electron subbands,
- the light hole subbands and
- the heavy hole subbands,

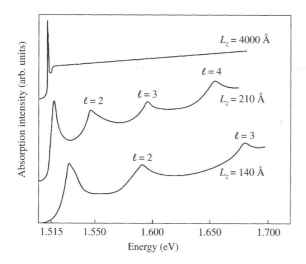

Fig. 7.5 Optical absorption of a series of GaAs-(Ga,Al)As quantum wells of differing widths L_z at 4.2 K. (After R. Dingle et al., *Phys. Rev. Lett.* **33**, 827 (1974).)

each subband within a set being labelled by a quantum number $i = 1, 2, 3\ldots$. Motion in the xy plane will be unrestricted; we therefore have a two-dimensional carrier system in the well.

7.3.4 Optical properties of quantum wells

The interband optical absorption or emission of a quantum well will be caused by transitions between hole and electron subbands. The selection rules are determined by the overlap between the electron and hole wavefunctions. Wavefunctions of the same quantum number (or quantum numbers differing by two) will have similar (spatial) shapes and the transition will be strong; wavefunctions whose quantum numbers differ by one will be of opposite symmetry, and so transitions between them will be weak. In summary, the selction rules are[2]

- $\Delta i = 0, 2$: strong, allowed transitions;
- $\Delta i = 1$: weak, 'forbidden' transitions.

[2] For a detailed derivation of these selection rules, see M. Fox, *Optical properties of solids*, (OUP, 2001).

As there are two sets of hole subbands, strong transitions will occur in pairs, e.g. (*i*th heavy-hole subband to *i*th electron subband) and (*i*th light-hole subband to *i*th electron subband). Figure 7.5 shows the optical absorption of a set of quantum wells of differing well width. The wells are of rather low quality, so that individual transitions are hard to distinguish. However, it is plain that

- the transition energies increase as the wells get narrower, due to the increased subband confinement energy;
- the absorption increases in steps, reflecting the step-like form of the electron–hole joint density of states in two dimensions;
- excitons are very prominent, because the confinement in the well enhances the exciton binding energy by holding the electron and hole closer together than in the bulk semiconductor.

Figure 7.6 shows the transmission of a set of much higher quality GaAs-(Ga,Al)As quantum wells. Unfortunately the detector used did not give a flat

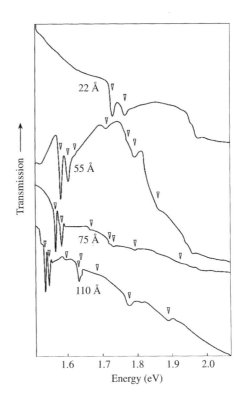

Fig. 7.6 Optical absorption of a series of GaAs-(Ga,Al)As quantum wells of differing widths L_z at 55 K. (Data from D.C. Rogers et al, *Phys. Rev. B* **34**, 4002 (1986).)

response so that the steps in absorption cannot be seen so clearly; however, the pairs of strong transitions can now be resolved. Note that the excitons involving light-holes are almost as strong as excitons involving heavy holes (cf. the bulk case (Fig. 6.8), where the latter dominate); this is because quirks of the valence-band bandstructure give the light-hole subbands quite a large in-plane (xy) effective mass, and therefore a large density of states, comparable to that of the heavy holes.

7.3.5 Use of quantum wells in opto-electronics

There are several reasons for the use of quantum wells in opto-electronic applications, e.g.

- the energy of the fundamental optical transition can be varied by varying the well width;
- the heavy- and light-hole degeneracy at the Brillouin-zone centre is broken, removing complications associated with scattering etc.;
- all of the transitions are excitonic (i.e. sharp features at a well-defined energy, rather than broad edges), even at 300 K;
- the well can be used to hold electrons and holes in close proximity, to encourage more efficient recombination in e.g. lasers and LEDs.[3]

Applications include the **Quantum-confined Stark effect modulator**, which is dealt with in detail in e.g. *Low-dimensional semiconductor structures*, by M.J. Kelly (Clarendon Press, Oxford, 1995) Section 18.8.

---

[3] See *Semiconductor physics*, by K. Seeger (Springer, Berlin, 1991) Section 13.2.

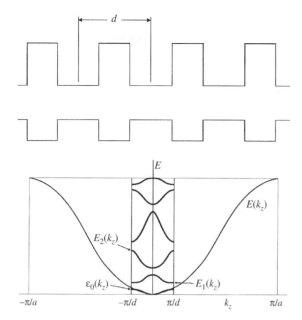

Fig. 7.7 Schematic of the band-edges in a superlattice.

Fig. 7.8 Minibands and minigaps in a superlattice (see Fig. 7.7). Here, π/a denotes the Brillouin-zone boundary of the underlying lattice; $E(k_z)$ represents the unperturbed, original bulk band. The effect of the superlattice, of period d, is to introduce new 'mini Brillouin zones' with boundaries at $\pm\pi/d$. The band is split into *minibands* with dispersion relationships labelled by $E_0(k_z)$, $E_1(k_z)$, $E_2(k_z)$ etc.

7.3.6 Superlattices

A **superlattice** contains a set of quantum wells which are sufficiently closely spaced for the carriers to tunnel between wells (see Fig. 7.7). The situation is analogous to the tight-binding or Kronig–Penney models of bandstructure; the subbands will broaden out to form **minibands** with **minigaps** between. One can also think of the periodicity of the superlattice introducing a new set of Brillouin-zone boundaries and hence energy gaps (see Fig. 7.8).

Superlattices have a variety of applications in **resonant tunnelling structures** and far-infrared detectors; we shall discuss more of the details of these devices in Chapter 12.[4]

[4] See also See *Low-dimensional semiconductor structures*, by M.J. Kelly (Clarendon Press, Oxford, 1995) Chapters 9 and 10, 17–19.

7.3.7 Type I and type II superlattices

Type I superlattices

Thus far, we have been using the system GaAs-(Ga,Al)As as an example in our discussion of quantum wells and superlattices.[5] In GaAs-(Ga,Al)As superlattices, the band offsets of the two materials are such that the electron and hole wells occur in the same material (GaAs).[6] Such an arrangement is known as a *Type I* superlattice; other examples include the systems (Ga,In)As-InP, (Ga,In)As-(Al,In)As and (Cd,Hg)Te-CdTe. Note that the band offsets vary between the different systems, so that it is possible to design superlattices in which, for example, the holes are strongly confined in deep wells, and the electrons are only weakly confined in very shallow wells.

[5] The various material systems mentioned in this Section are described in more detail in *Low-dimensional semiconductor structures*, by M.J. Kelly (Clarendon Press, Oxford, 1995).

[6] By implication, the *barriers* for electrons and holes are both provided by the (Ga,Al)As.

Type II superlattices: uncrossed bands

In a Type II superlattice, the electron wells occur in one material and the hole wells occur in the other. Type II superlattices can be divided into two further

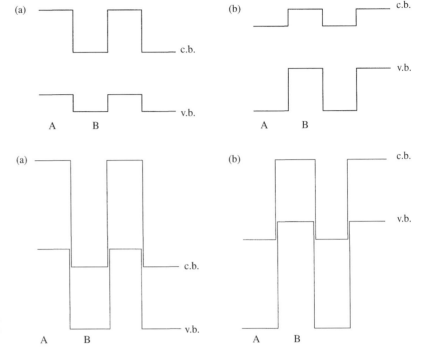

Fig. 7.9 Schematic band line-up in type II superlattices without crossed gaps (c.b. = conduction band-edge; v.b. = valence band-edge). Two examples are shown to illustrate the effect that varying the band offsets can have; (a) has a deep electron well in material B; (b) has a deep hole well in material B.

Fig. 7.10 Schematic band line-up in Type II superlattices with crossed gaps (c.b. = conduction band-edge; v.b. = valence band-edge). Two examples are shown to illustrate the effect that varying the band offsets can have; (a) is similar to the line-up in the InAs-(Ga,In)Sb system.

[7] Note that it is in principle possible to make a Type II superlattice using two semiconductors with equal band gaps.

classes according to whether or not the electron wells are at a higher energy than the hole wells. The simplest situation occurs in superlattices where the bottom of the electron wells is at a higher energy than the top of the hole wells (the bands are said to be uncrossed). Fig. 7.9 shows two possible arrangements of the bands. An interesting aspect of such systems is that the hole and electron states are now somewhat spatially separated; in the case of wide layers, the electron subband wavefunctions will be mainly in one material and those of the holes in the other. In the case of superlattices, the electron and hole miniband wavefunctions will have greatest amplitude in different layers. This spatial separation will greatly modify the strength of the optical transitions between hole and electron states.

Examples of Type II superlattices without crossed gaps include Si-Ge and (Ga,In)As-GaAs.[7]

Type III superlattices: Type II superlattices with crossed gaps

In Type II systems with a crossed gap, the electron wells are at a lower energy than the hole wells (these are also known as 'Type III superlattices'). The situation is shown in Fig. 7.10; the band-gap of material B lies entirely outside the band gap of material A. The InAs-(Ga,In)Sb system is an example of a Type II crossed-gap superlattice. (Ga,In)Sb has the larger band gap, and its conduction band lies at a higher energy than that of InAs; the band line-up is similar to that in Fig. 7.10(a).

InAs-(Ga,In)Sb superlattices are of some interest because a semiconductor-to-semimetal transition can be engineered by varying the well widths. The energetic separation of the electon states (in the InAs) and the hole states

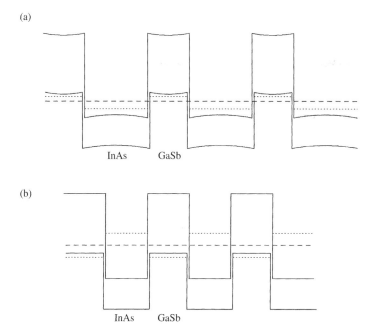

Fig. 7.11 Band profile of InAs-(Ga,In)Sb superlattices, showing the lowest electron and hole states in the InAs and (Ga,In)Sb respectively (dotted lines). The chemical potential is shown as a dashed line. (a) A semimetallic arrangement (wide InAs layers), in which transfer of electrons from (Ga,In)Sb to InAs occurs, leaving behind holes. (b) A semiconducting arrangement (narrow InAs layers) in which the lowest hole state (in (Ga,In)Sb) is lower than the lowest electron state (in InAs), with the chemical potential in between.

(in the (Ga,In)Sb) is mainly determined by the InAs well width, since the InAs conduction-band effective mass is much smaller than that of the holes in (Ga,In)Sb. If the layers are wide (see Fig. 7.11(a)), the electron states in InAs are below the hole states in (Ga,In)Sb; electron transfer occurs from the valence band of (Ga,In)Sb (leaving holes behind) to the conduction-band well of InAs. The system is thus an intrinsic semimetal;[8] electrons and holes are simultaneously present without doping.

As the thickness of the layers decreases, the confinement energies increase until the electron states in InAs are above the hole states in (Ga,In)Sb (Fig. 7.11(b)), resulting in a semiconducting system. The cross-over occurs at an InAs layer thickness of about 85 Å. The semiconducting superlattices are of potential interest as far-infrared photodetectors.

[8] It has equal numbers of electrons and holes, and both are degenerate systems (i.e. they possess Fermi surfaces).

7.3.8 Heterojunctions and modulation doping

A heterojunction is a single junction between two layers of different semiconductors. A typical example is the GaAs-(Ga,Al)As heterojunction. First, a thick layer of undoped GaAs is grown, followed by a few tens of nm of (Ga,Al)As (the *spacer layer*). After this a section of heavily-doped ($n \sim 10^{18}$ cm^{-3}) (Ga,Al)As is deposited, followed by more undoped (Ga,Al)As (not shown). The final band arrangement is shown in Fig. 7.12. The combination of the conduction band offset at the interface and the pinning of the chemical potential inside the GaAs and (Ga,Al)As layers results in a one-dimensional, approximately triangular potential well containing electrons which have 'dropped off' the donors in the (Ga,Al)As. As in the case of the quantum well, the electrons' z-direction motion is quantised into subbands; however, the xy motion is unconstrained, so that a degenerate (metallic) two-dimensional electron system results.

74 *Bandstructure engineering*

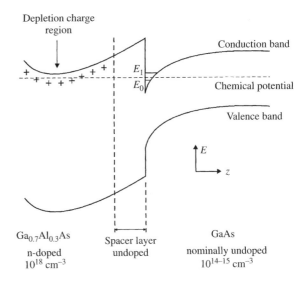

Fig. 7.12 The band-edges in a modulation-doped GaAs-(Ga,Al)As heterojunction.

The spacer layer separates the ionised donors from the electrons, dramatically reducing the scattering; low-temperature mobilities $\mu_c \sim 10^7\,\mathrm{cm^2 V^{-1} s^{-1}}$ are possible using this technique, which is known as **modulation doping** (because the dopant concentration in the (Ga,Al)As is modulated).

Applications of heterojunctions include the **High-electron-mobility transistor** (HEMT), a field-effect transistor with a heterojunction as active layer.[9] Most experiments on the **quantum Hall effect** are carried out using heterojunctions; we shall study this effect in detail in Chapter 11.

Other applications of modulation doping include the fabrication of so-called δ-doping layers, in which a very heavy dose of dopant atoms is deposited in a small width (∼ a few monolayers). In effect, this forms a two-dimensional impurity band, a very thin metallic layer within the host semiconductor. Such layers are finding applications as metallic gates and contacts *within* complex, layered semiconductor structures.

[9] See *Low-dimensional semiconductor structures*, by M.J. Kelly (Clarendon Press, Oxford, 1995) Chapter 16.

7.3.9 The envelope-function approximation

The **envelope-function approximation** is often used to calculate the electronic energy levels in heterostructures, such as GaAs-(Ga,Al)As quantum wells, which are made up from two or more similar semiconductors. Because the two semiconductors involved are very similar from both chemical and crystallographic points of view, it is assumed that the rapidly-oscillating part of the Bloch function (the part which has the periodicity of the lattice) is the same in both materials; only the *envelopes* ϕ of the Bloch functions differ in the two semiconductors.

As an example, consider the GaAs-(Ga,Al)As quantum well shown in Fig. 7.4; let a typical envelope function in the GaAs well be $\phi_A(z)$ and the corresponding envelope function in the (Ga,Al)As barriers be $\phi_B(z)$. The boundary conditions in the envelope function approximation at the interfaces (i.e. at $z = \pm \frac{a}{2}$) are

$$\phi_A = \phi_B \tag{7.2}$$

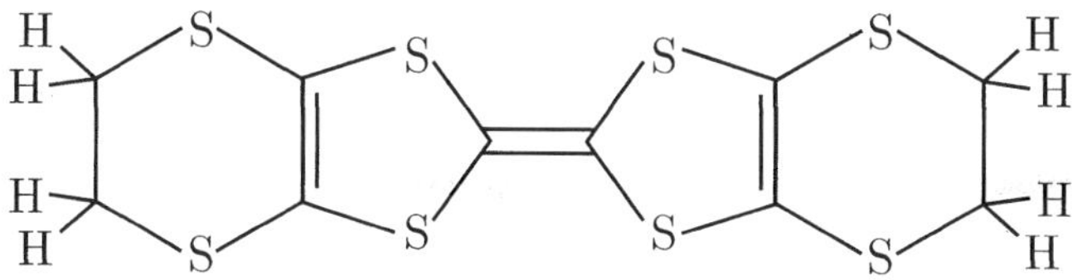

Fig. 7.13 The bisethylenedithiotetrathioful-valene (BEDT-TTF) molecule.

and
$$\frac{1}{m_A^*}\frac{\mathrm{d}\phi_A}{\mathrm{d}z} = \frac{1}{m_B^*}\frac{\mathrm{d}\phi_B}{\mathrm{d}z}, \tag{7.3}$$

where m_A^* and m_B^* are the effective masses in the well and barrier respectively. Equation 7.3 ensures that the probability-density flux is conserved.

By considering envelope functions such as $\cos(kz)$ and $\sin(kz)$ in the well and evanescent waves $e^{\pm \kappa z}$ in the barriers, the subband energies can be found. The exercises contain an example of this technique, which can easily be extended to more complex structures such as superlattices.

7.4 Band engineering using organic molecules

7.4.1 Introduction

In the earlier sections of this chapter, the problem of designing 'tailor-made' bandstructure was approached by using multilayers of different semiconductors to form superlattices and quantum wells. In this section, we shall examine another approach which involves the use of organic molecules to 'build' systems with easily controllable electronic properties, so-called organic molecular metals. The relationship between the crystal structures of these materials and their bandstructure may be easily understood using the tight-binding model, and so much of this section represents a useful revision of Chapter 4 and the other chapters on bandstructure.

Organic molecular metals are the subject of a considerable research effort world-wide; the techniques involving high magnetic fields which are described later in Chapter 8 are extensively applied in the study of the Fermi surfaces of organic molecular metals, and one of these materials will be used as a case-study in that chapter.

7.4.2 Molecular building blocks

In our discussion of the tight-binding model (Chapter 4), we saw that atomic orbitals were used to build up the Bloch functions which give rise to the bandstructure. Similarly, the bandstructure of organic molecular metals is derived from **molecular** orbitals. Bisethylenedithiotetrathiofulvalene (also known as BEDT-TTF or ET; Fig. 7.13) is typical of the molecules used as 'bandstructure building-blocks' in organic molecular metals.[10] It is roughly flat, so that it can be stacked in a variety of arrangements in a solid, and it is surrounded by voluminous molecular orbitals. In order to form the bandstructure, BEDT-TTF molecules must then be stacked next to each other so that the molecular orbitals overlap; crudely one might say that this enables the electrons to transfer from molecule to molecule.

[10] Other molecules used for this purpose are described in *Organic superconductors* by T. Ishiguo, K. Yamaji and G. Saito (Springer-Verlag, Berlin 1998) and *Fermi surfaces of low-dimensional organic metals and superconductors* by J. Wosnitza (Springer-Verlag, Berlin 1996).

76 Bandstructure engineering

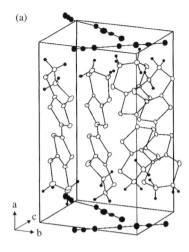

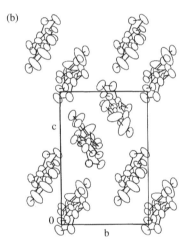

Fig. 7.14 Structure of the BEDT-TTF charge-transfer salt κ-(BEDT-TTF)$_2$Cu(NCS)$_2$. (a) Side view of molecular arrangement; the BEDT-TTF molecules pack in planes separated by layers of the smaller Cu(NCS)$_2$ anions. (b) View downwards onto the BEDT-TTF planes, showing that the BEDT-TTF molecules are packed closely togther, allowing substantial overlap of the molecular orbitals. (Based on H. Urayama et al., Chem. Lett. **1988**, 463.)

The disposition of the bandstructure-forming molecules in an ordered arrangement is usually accomplished by making a **charge-transfer salt**. In a charge-transfer salt, a number j of BEDT-TTF molecules will jointly donate an electron to a second type of molecule (or collection of molecules) which we label X, to form the compound (BEDT-TTF)$_j X$; owing to its negative charge, X is known as the **anion**. The transfer of charge serves to bind the charge-transfer salt together (in a manner analogous to ionic bonding) and also leaves behind a hole, jointly shared between the j BEDT-TTF molecules. This means that the bands formed by the overlap of the BEDT-TTF molecules will be **partially filled**; i.e. the resulting salt will be metallic.

Figure 7.14 shows the molecular arrangements in the BEDT-TTF charge-transfer salt κ-(BEDT-TTF)$_2$Cu(NCS)$_2$. (Here the κ denotes the packing arrangement of the BEDT-TTF molecules; often several different packing arrangements can be achieved with one particular anion.) The BEDT-TTF molecules are packed into layers, separated by layers of the Cu(NCS)$_2$ anion molecules. Within the BEDT-TTF layers, the molecules are in close proximity to each other, allowing substantial overlap of the molecular orbitals; in the language of Section 4.2.3, we can say that the **transfer integrals** for hopping of carriers between BEDT-TTF molecules will be relatively large within the BEDT-TTF planes. Conversely, in the direction perpendicular to the BEDT-TTF planes, the BEDT-TTF molecules are well separated from each other; the transfer integrals will be much smaller in this direction. This extreme anisotropy, resulting from a layered structure, is typical of most organic molecular metals; it results in bandstructures and electronic properties which for many purposes can be considered to be two-dimensional.

Figure 7.15 shows another BEDT-TTF charge-transfer salt, β-(BEDT-TTF)$_2$I$_3$ (again β denotes the packing arrangement of the BEDT-TTF molecules). As in the case of κ-(BEDT-TTF)$_2$Cu(NCS)$_2$ (Fig. 7.14), the BEDT-TTF molecules pack in planes separated by layers of the smaller anions, in this case I$_3$ molecules. However, the packing arrangement of the BEDT-TTF molecules is considerably different from that in κ-(BEDT-TTF)$_2$Cu(NCS)$_2$.

A comparison of Fig.s 7.14 and 7.15 shows that charge-transfer salts are an extremely flexible system with which to study the physics of band formation; by changing the anion, one can make the BEDT-TTF molecules pack in different arrangements. As the BEDT-TTF molecules are long and flat (unlike single atoms, which are far more like spheres), the transfer integrals will depend strongly on the way in which the BEDT-TTF molecules are arranged with respect to each other; this will be reflected in the *shape* of the resulting bands.

Far more subtle changes are also possible. β-(BEDT-TTF)$_2$I$_3$ is a member of the so-called β-phase BEDT-TTF salts, which have the generic formula β-(BEDT-TTF)$_2 X$, where the anion molecule X can be I$_3$, IBr$_2$, AuI$_2$ etc. The different possible anions have slightly different lengths; therefore the size of the unit cell can be varied by using different anions (see Fig. 7.15). If a longer anion (e.g. $X = $ I$_3$) is used, the BEDT-TTF molecules will be further apart, so that the transfer integrals will be smaller; the resulting bands will be narrower and the effective masses larger. Conversely, if a shorter anion (e.g. $X = $ IBr$_2$) is used, the BEDT-TTF molecules will be closer, so that the transfer integrals will be larger; the resulting bands will be wider and the effective masses smaller.

7.4 Band engineering using organic molecules 77

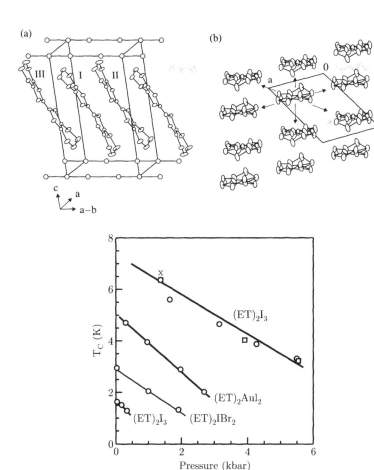

Fig. 7.15 Structure of the BEDT-TTF charge-transfer salt β-(BEDT-TTF)$_2$I$_3$. (a) Side view of molecular arrangement. As in Fig. 7.14, the BEDT-TTF molecules pack in planes separated by layers of the smaller anions; in this case the anions are I$_3$ molecules. (b) View downwards onto the BEDT-TTF planes, showing that the BEDT-TTF molecules are packed closely togther, allowing substantial overlap of the molecular orbitals. However, the packing arrangement of the BEDT-TTF molecules is considerably different from that in κ-(BEDT-TTF)$_2$Cu(NCS)$_2$ (see Fig. 7.14). (Based on H. Kuroda *et al.*, *Synth. Met.* **27**, A491 (1988).)

Fig. 7.16 Superconducting critical temperature T_c versus hydrostatic pressure for three β-phase BEDT-TTF salts (BEDT-TTF has been abbreviated to ET to save space). Decreasing the unit cell size, either by using a shorter anion or by increasing the pressure, reduces T_c. (Note that β-(BEDT-TTF)$_2$I$_3$ undergoes a structural phase transition at about 0.6 kbar.) (Based on data compiled in *Organic Superconductors*, by T. Ishiguro *et al.* (Springer, 1998).)

Thus, different anions can be employed to optimise a particular physical property; an example of this is shown in Fig. 7.16, which shows the relationship between superconducting critical temperature T_c (see Appendix H) and hydrostatic pressure for three β-phase salts. It can be seen that increasing the pressure (which decreases the unit cell size, thereby increasing the width of the bands) lowers T_c; substitution of a larger anion increases the unit cell size, countering the effect. We shall see why the bandwidth affects T_c in Chapter 8.

7.4.3 Typical Fermi surfaces

A section through the first Brillouin zone and Fermi surface of κ-(BEDT-TTF)$_2$Cu(NCS)$_2$ is shown in Fig. 7.17. The section shown there is parallel to the highly-conducting planes; as has already been remarked, the electronic properties of the material are largely two-dimensional. The Fermi surface may be understood by examining Fig. 7.14(b). Note that the BEDT-TTF molecules are packed in pairs called **dimers**; if we consider each dimer as a unit, it will be seen that it is surrounded by four other dimers. We thus expect the bandstructure to be fairly isotropic in the plane, because there will be substantial transfer integrals in these four directions. The unit cell contains two

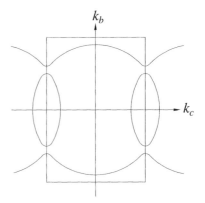

Fig. 7.17 Brillouin zone and Fermi surface of κ-(BEDT-TTF)$_2$Cu(NCS)$_2$, showing the open, quasi-one-dimensional sections, and the closed, quasi-two-dimensional pocket. (J.M. Caulfield *et al.*, *J. Phys.: Condens. Matter*, **6**, 2911 (1994).)

dimers, each of which contributes a hole (in the charge-transfer process, two BEDT-TTF molecules jointly donate one electron to the anion). Thus, to a first approximation, the Fermi surface for holes might be expected to be roughly circular with the same area as the Brillouin zone (as there are two holes per unit cell). As can be seen from Fig. 7.17, the Fermi surface is *roughly* like this; however, the Fermi surface intersects the Brillouin zone boundaries in the **c** direction, so that band gaps open up. The Fermi surface thus splits into open (electron-like) sections running down two of the Brillouin-zone edges and a closed hole pocket straddling the other.

7.4.4 A note on the effective dimensionality of Fermi-surface sections

The Fermi surface of κ-(BEDT-TTF)$_2$Cu(NCS)$_2$ has been shown to consist of both open and closed sections (see Fig. 7.17); it is customary to label such sections 'quasi-one-dimensional' and 'quasi-two-dimensional' respectively. The names arise because the group velocity **v** of the electrons is given by eqn 5.5

$$\hbar \mathbf{v} = \nabla_\mathbf{k} E(\mathbf{k}).$$

The Fermi surface is a surface of constant energy; eqn 5.5 shows that the velocities of electrons at the Fermi surface will be directed perpendicular to it. Therefore, referring to Fig. 7.17, electrons on the closed Fermi-surface pocket can possess velocities which point in any direction in the (k_b, k_c) plane; they have freedom of movement in two dimensions and are said to be **quasi-two-dimensional**. By contrast, electrons on the open sections have velocities predominently directed parallel to k_b and are **quasi-one-dimensional**.

Quasi-two-dimensional and quasi-one-dimensional Fermi-surface sections are observed in many layered materials, and so it is worth getting to know how to identify them.

7.5 Layered conducting oxides

Oxides of transition metal ions such as Mn, Fe, Ni, V, Ru and Cu have received considerable attention because of their electronic properties. In this Section I

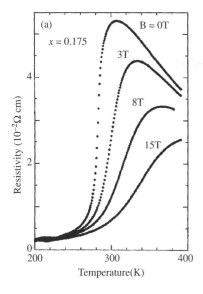 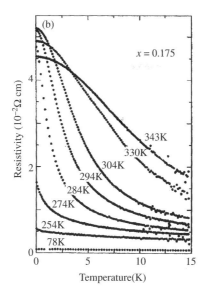

Fig. 7.18 (a) The resistivity of $La_{1-x}Sr_xMnO_3$ ($x = 0.175$) as a function of temperature for various different magnetic fields. (b) The resistivity of $La_{1-x}Sr_xMnO_3$ ($x = 0.175$) as a function of field for various different temperatures. (After Y. Tokura et al., J. Phys. Soc. Jpn. **63**, 3931 (1994).)

shall describe the oxides of Mn as an example of this interest, and show how the introduction of layered crystal structures can be used in attempts to optimise desirable properties.

Figure 7.18 shows the electrical resistivity of $La_{1-x}Sr_xMnO_3$ ($x = 0.175$) as a function of temperature for various different magnetic fields. At first, it might seem surprising that such a compound should conduct electricity; however, this occurs because the Mn ion can exhibit what is called **mixed valence**. In other words, the Mn ion can exist as *either* Mn^{3+} *or* Mn^{4+}. The parent compound, $LaMnO_3$, is an insulator in which all of the Mn ions are in the Mn^{3+} state (La is a trivalent ion, La^{3+}). In $La_{1-x}Sr_xMnO_3$, trivalent La ions are replaced by divalent Sr^{2+} ions, resulting in a shortage of electrons (or a surplus of holes); hence a fraction x of the Mn ions will lose an extra electron, so that they will end up in the Mn^{4+} state.

The important region of the bandstructure for our purposes involves two quantum levels associated with the Mn ions and their surrounding O ions. The lower level is highly localised on the Mn ions, and is known as the t_{2g} level. **Exchange interactions**[11] dictate that three spin-polarised electrons can be accommodated in this level; the subsequent electron should be placed in the level above (the e_g level), and should also have the same spin polarisation as the t_{2g} electrons. Thus, Mn^{3+} ions have three electrons in the t_{2g} level and one electron in the e_g level, whereas the Mn^{4+} ions have empty e_g levels.

Whereas the t_{2g} levels are very tightly bound (effectively rendering the electrons therein immobile), the e_g levels overlap to form a narrow band, via which electrons can hop from Mn to Mn; oxides in which there is a mixture of Mn^{3+} and Mn^{4+} therefore have a partly filled band of mobile electrons which can lead to electrical conduction. The spectacular dependence of the resistivity of $La_{1-x}Sr_xMnO_3$ on magnetic field and temperature occurs because of another manifestation of the exchange interactions, known as **double exchange**,[12] which strongly suppresses hopping of an e_g electron from an Mn^{3+} ion to an Mn^{4+} ion *unless* both have their t_{2g} spins polarised in

[11] For the present purposes, these interactions should just be accepted as fact, as this book does not seek to present a detailed description of magnetism in solids. For a full discussion of exchange interactions in solids, see *Magnetism in condensed matter*, by S. Blundell (OUP, 2001).

[12] See Chapter 4 of *Magnetism in condensed matter*, by S. Blundell (OUP, 2001).

Fig. 7.19 Ruddlesden–Popper oxides (general formula $A_{n+1}B_nO_{3n+1}$) with $n = 1, 2, \infty$. The parameter n represents the number of layers of vertex shared BO_6 octahedra stacked along the crystallographic direction [001]. Here the A ions are grey, the B ions black and the O ions white. (Figure courtesy of Steve Blundell, Oxford University).

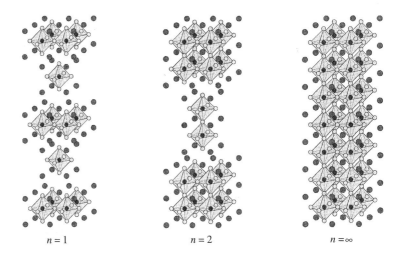

$n = 1$ $n = 2$ $n = \infty$

the same direction. $La_{1-x}Sr_xMnO_3$ ($x = 0.175$) undergoes a transition into a ferromagnetic state in which all of the t_{2g} electron spins are aligned at about 300 K; hence, the e_g electrons will be able to hop with ease and a relatively low resistance will result; this is the explanation of the large drop in resistivity at around 300 K in Fig. 7.18.

At temperatures above 300 K, hopping of the e_g electrons will be suppressed, as the spin system is unpolarised. However, the application of a magnetic field will act to align the spins of all of the electrons on the Mn ions, so that hopping will become easier, leading to the spectacular decreases in resistivity seen in Fig. 7.18. Such very large decreases in resistivity caused by the application of a magnetic field have become known as **colossal magnetoresistance**;[13] materials exhibiting colossal magnetoresistance have great potential as, for example, resistive (as opposed to inductive) read heads for magnetic recording media.

The performance of oxides of Mn exhibiting colossal magnetoresistance depends on the interactions which cause the magnetic order, and the overlap of the Mn ions with their neighbours which determines the properties of the e_g band. One way of modifying these effects involves changing the effective dimensionality of these materials. $La_{1-x}Sr_xMnO_3$ is a member of a series of compounds known as **Ruddlesden–Popper oxides**, which possess a general formula $A_{n+1}B_nO_{3n+1}$; here n represents the number of layers of vertex shared BO_6 octahedra stacked along the crystallographic direction [001]. Examples with $n = 1$ (A_2BO_4), 2 ($A_3B_2O_7$) and ∞ (ABO_3) are shown in Fig. 7.19; it will be seen that the structures with $n = 1, 2$ are quasi-two-dimensional, in that the BO_6 octahedra which are responsible for the e_g band are planes which are rather well separated in the [001] direction. By contrast, $La_{1-x}Sr_xMnO_3$ corresponds to the case $n = \infty$, and possesses much more isotropic (i.e. more 'three-dimensional') bandstructure and magnetic interactions.[14]

Much work continues on the effect of varying the effective dimensionality of oxide materials in an attempt to optimise desirable properties. Amongst studies of this kind which have yielded great gains is the development of the 'high T_c' cuprate superconductors. Values of T_c well in excess of 100 K have been

---

[13] A much smaller magnetoresistance is seen in very many electrically conductive materials, but its origin is rather different; it is usually caused by modification of the real-space paths of the carriers due to the Lorentz force. This form of magnetoresistance will be dealt with in Chapter 10.

[14] The $n = \infty$ case is often referred to as the **perovskite** structure.

achieved through the production of structures in which the electronically active layers containing Cu and O have been progressively separated by including more intermediate layers of 'packing'. A detailed discussion of this is beyond the scope of this book, but an excellent summary of the development of layered cuprate superconductors is given in Chapter 7 of *Superconductivity*, by C.P. Poole, H.A. Farach and R.J. Creswick (Academic Press, San Diego, 1995) and in *Superconductivity*, by J. Annett (OUP, 2002).

7.6 The Peierls transition

Many of the layered substances (organic molecular solids, oxides) discussed in this chapter have Fermi-surface sections which are quasi-one-dimensional (see Section 7.4.4 for an explanation of this terminology). Such substances are unstable with respect to what are known as **Peierls transitions**, in which a periodic modulation of the crystal structure results. Figure 7.20 describes

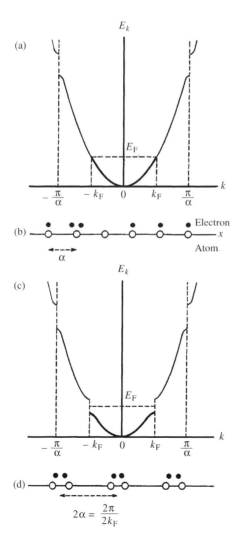

Fig. 7.20 (a) A simple nearly-free-electron band in a monovalent one-dimensional material; band gaps exist at the Brillouin-zone boundary and the Fermi surface (a pair of points) has a width equal to half that of the Brillouin zone. (b) The equivalent of (a) plotted in r-space, showing (on average) one electron (black dot) per atom (hollow point); an electron has been 'caught in the act' of hopping from one atom to the next. (c) The band after the Peierls distortion shown in (d). The crystal has distorted, doubling the r-space unit-cell size, and halving the Brillouin-zone size. As new zone boundaries exist at $\pm k_F$, a band-gap opens up there, lowering the electronic energy, and turning the substance into an insulator, with two electrons per unit cell.

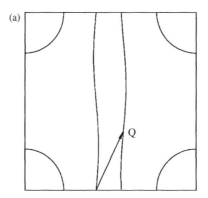 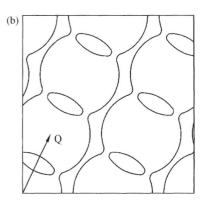

Fig. 7.21 (a) A Fermi surface (consisting of a pair of quasi-one-dimensional sheets plus a quasi-two-dimensional cylinder) which is prone to spin- or charge-density wave formation. The nesting vector Q which maps the quasi-one-dimensional Fermi-surface sections onto each other (so that a band-gap opens along their whole length) is shown as an arrow. (b) Rearrangement of the Fermi surface in (a) caused by a spin- or charge-density wave (the arrow indicates the nesting vector). A new k-space periodicity and new Brillouin zone have been introduced; the remaining section of Fermi surface has been 'cut and pasted' with this periodicity, resulting in a much altered bandstructure.

this schematically in a one-dimensional material. Figure 7.20(a) displays a nearly-free-electron band in a monovalent one-dimensional material. As we saw in Chapter 3, band gaps exist at the Brillouin-zone boundary at $\pm\frac{\pi}{a}$ and the Fermi surface (a pair of points) is half the length of the Brillouin zone (i.e. $k_F = \pm\frac{\pi}{2a}$). The situation in r-space is shown in Fig. 7.20(b); on average there is one electron per atom, and electrons are free to move along the chain of atoms, which are spaced by a. It turns out that in this situation, a very simple distortion of the r-space lattice can lower the energy of the electron system. Such a distortion is shown in Fig. 7.20(d); adjacent atoms have moved together to form **dimers**; the unit-cell length has doubled to $2a$, so that the Brillouin zone has halved in size (see Fig. 7.20(c); the new boundaries are at $\pm k_F = \pm\frac{\pi}{2a}$). A band-gap opens up at the new Brillouin-zone boundaries, lowering the electronic energy, and turning the substance into an insulator. This sort of distortion will be able to occur as long as the electrons can interact with the lattice.

The distortion has resulted in electronic charge piling up on the dimers; the resulting state is known as a **charge-density wave**, as the electron density (once much more uniform) has become modulated. A similar lowering of energy can be obtained by doubling the unit cell with a periodic modulation of the electron spin; this is known as a **spin-density wave**.[15]

[15] See Chapter 8 of *Magnetism in condensed matter*, by S. Blundell (OUP, 2001) and Chapter 2 of *One-dimensional conductors*, by S. Kagoshima, H. Nagasawa and T. Sambongi (Springer-Verlag, Berlin, 1988).

Modulations of this type may be described by a series of terms $\propto \cos(Qx)$, where the wavevector $Q(=2k_F)$ is known in the trade as a **nesting vector**.

It turns out that charge- and spin-density waves are likely to occur in non-one-dimensional materials whenever a single nesting vector can cause large areas of Fermi surface to map onto each other; in r-space terms, this means introducing a distortion with a periodicity which creates band gaps over large sections of Fermi surface, so that a large fraction of the electrons can lower their energy. Fig. 7.21 illustrates such a case.

Peierls transitions are remarkably common in organic molecular metals and oxides and are even observed in a few elemental metals such as chromium.

Further reading

A more detailed treatment of semiconductor heterostructures is found in *Quantum semiconductor structures* by Claude Weisbuch and Borge Winter (Academic Press, San Diego, 1991) and in *Low-dimensional semiconductor structures*, by M.J. Kelly (Clarendon Press, Oxford, 1995), Chapters 2, 9 and 10; a complementary (simple) summary is given in *Solid state physics*, by G. Burns (Academic Press, Boston, 1995), Sections 18.1–18.6. The book *Quantum wells, wires and dots: theoretical and computational physics*, by Paul Harrison (Wiley, Chichester, 1999) is an excellent point of departure if you want to carry out calculations and simulations. An good review of the envelope-function model is given in *Structure de bandes, niveaux d'impureté et effets Stark dans les super-réseaux* by G. Bastard, *Acta Electronica* **25**, 147–161 (1983). Optical properties of semiconductor structures are tackled in depth in *Optical properties of solids*, by M. Fox (OUP, 2001).

Reviews of organic molecular metals are given in *Organic superconductors* by T. Ishiguo, K. Yamaji and G. Saito (Springer-Verlag, Berlin 1998), *Fermi surfaces of low-dimensional organic metals and superconductors* by J. Wosnitza (Springer-Verlag, Berlin 1996) and by J. Singleton in *Reports on Progress in Physics* **63**, 1111 (2000).

Electrically conductive oxides are discussed in *Transition metal oxides*, by P.A. Cox (OUP, 1995) and in *Metal-insulator transitions*, by N.F. Mott (Taylor and Francis, London, 1990 (second edition)). Relatively recent reviews of colossal magnetoresistance have been provided by A.P. Ramirez (*J. Phys.: Condens. Matter* **9**, 8171 (1997)) and C.N.R. Rao (*J. Mater. Chem.* **9**, 1 (1999)).

An excellent review of Peierls transitions (at a reasonably simple level) is given in Chapter 2 of *One-dimensional conductors*, by S. Kagoshima, H. Nagasawa and T. Sambongi (Springer-Verlag, Berlin, 1988). See also G. Grüner, *Rev. Mod. Phys.* **60**, 1129 (1988) and *Rev. Mod. Phys.* **66**, 1 (1994).

Exercises

The problems in this chapter explore the envelope-function approximation, using the GaAs-(Ga,Al)As quantum well shown in Fig. 7.22 as an example. The GaAs layer forms a one-dimensional potential well, sandwiched between $Ga_{0.65}Al_{0.35}As$ barriers. The energy gaps of GaAs and $Ga_{0.65}Al_{0.35}As$ are $E_{g1} = 1.52$ eV and $E_{g2} = 1.97$ eV respectively and you can assume that $\frac{\Delta E_c}{\Delta E_v} \approx 1.5$. The GaAs layer is 100 Å thick. Effective masses in GaAs and (Ga,Al)As are given in Table 7.1.

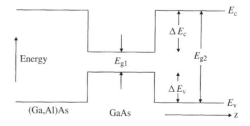

Fig. 7.22 Schematic of a GaAs-(Ga,Al)As quantum well.

Table 7.1 Effective masses (in units m_e) of GaAs and $Ga_{0.65}Al_{0.35}As$.

Material	Electron	Light hole	Heavy hole
GaAs	0.07	0.085	0.45
(Ga,Al)As	0.091	0.11	0.5

(7.1) Let a typical envelope function in the GaAs well be $\phi_A(z)$ and the corresponding envelope function in the (Ga,Al)As barriers be $\phi_B(z)$. The boundary conditions in the envelope-function approximation at the interfaces (i.e. at $z = \pm\frac{a}{2}$) are

$$\phi_A = \phi_B \quad (7.4)$$

and

$$\frac{1}{m_A^*}\frac{d\phi_A}{dz} = \frac{1}{m_B^*}\frac{d\phi_B}{dz}, \quad (7.5)$$

where m_A^* and m_B^* are the effective masses in the well and barrier respectively. Explain why envelope functions such as $\cos(kz)$ and $\sin(kz)$ are appropriate for the well and $e^{\pm\kappa z}$ for the barriers, and show that the bound states in the well are given by solutions of the equations

$$\tan(\frac{ka}{2}) = -\frac{m_B^* k}{m_A^* \kappa} \quad (7.6)$$

and

$$\tan(\frac{ka}{2}) = +\frac{m_A^* \kappa}{m_B^* k}. \quad (7.7)$$

(7.2) Using eqns 7.6 and 7.7, and the well parameters given above, find the energies of the bound states for the electrons, heavy holes and light holes. (IMPORTANT HINT: Equations 7.6 and 7.7 cannot be solved analytically. The problem can either be done *numerically*, by writing a simple program in C, PASCAL or some other language or *graphically*, using your favourite spreadsheet.)

(7.3) Label the state closest to the bottom of the well $i = 1$ and the next $i = 2$ etc. for each type of carrier (remember that the 'lowest' hole state is that closest to the GaAs valence band-edge). Sketch the low-temperature optical transmission of the well as a function of frequency ν in the range $E_{g1} \leq h\nu \leq E_{g2}$, assuming that the dominant transitions have $\Delta i = 0$.

Measurement of bandstructure

8.1 Introduction

This chapter will describe a selection of the most common techniques used to measure the bandstructures of solids. Many of these techniques involve the use of magnetic fields to partially quantise the electronic motion; this is known as Landau quantisation or cyclotron motion. I shall therefore treat the motion of band electrons in some detail before giving brief details of the techniques themselves.

8.1	Introduction	85
8.2	Lorentz force and orbits	85
8.3	The introduction of quantum mechanics	87
8.4	Quantum oscillatory phenomena	91
8.5	Cyclotron resonance	97
8.6	Interband magneto-optics in semiconductors	100
8.7	Other techniques	102
8.8	Some case studies	105
8.9	Quasiparticles: interactions between electrons	112

8.2 Lorentz force and orbits

8.2.1 General considerations

Equation 5.9 may be used to calculate the effect of an external magnetic field $\mathbf{B}$ on a band electron, i.e.

$$\hbar \frac{d\mathbf{k}}{dt} = -e\mathbf{v} \times \mathbf{B}, \quad (8.1)$$

where the right-hand side is the well-known Lorentz force. Equation 8.1 implies that

- the component of $\mathbf{k}$ parallel to $\mathbf{B}$ is constant;
- as $\frac{d\mathbf{k}}{dt}$ is perpendicular to $\mathbf{v}$ (defined property of the $\times$ operation) and as

$$\mathbf{v} = \frac{1}{\hbar} \nabla_{\mathbf{k}} E(\mathbf{k}) \quad (8.2)$$

(see eqn 5.5), $\frac{d\mathbf{k}}{dt}$ is perpendicular to $\nabla_{\mathbf{k}} E(\mathbf{k})$. This means that the electron path (orbit) is one of constant energy. To see this, consider $\frac{d\mathbf{k}}{dt} \cdot \nabla_{\mathbf{k}} E(\mathbf{k}) = 0$ (because the two are perpendicular); writing this in components gives $\frac{\partial k_x}{\partial t}\frac{\partial E}{\partial k_x} + \frac{\partial k_y}{\partial t}\frac{\partial E}{\partial k_y} + \frac{\partial k_z}{\partial t}\frac{\partial E}{\partial k_z} \equiv \frac{dE}{dt}$ from the chain rule. Hence, $\frac{dE}{dt} = 0$.

In k-space, the possible electron orbits are therefore described by the intersections of surfaces of constant energy with planes perpendicular to $\mathbf{B}$ (see also Exercise 8.2).

8.2.2 The cyclotron frequency

Consider a section of k-space orbit of constant energy E in a plane perpendicular to $\mathbf{B}$ (see Fig. 8.1). The time $t_2 - t_1$ taken to traverse the part of the orbit between $\mathbf{k}_1$ and $\mathbf{k}_2$ is

$$t_2 - t_1 = \int_{t_1}^{t_2} dt = \int_{\mathbf{k}_1}^{\mathbf{k}_2} \frac{dk}{|\dot{\mathbf{k}}|}, \qquad (8.3)$$

where $\dot{\mathbf{k}} = \frac{d\mathbf{k}}{dt}$. Equations 8.1 and 8.2 can be used to obtain $\dot{\mathbf{k}}$, so that eqn 8.3 becomes

$$t_2 - t_1 = \frac{\hbar^2}{eB} \int_{\mathbf{k}_1}^{\mathbf{k}_2} \frac{dk}{|\nabla_{\mathbf{k}\perp} E|}, \qquad (8.4)$$

where $\nabla_{\mathbf{k}\perp} E$ is the component of $\nabla_{\mathbf{k}} E$ perpendicular to the field.

The quantity $\nabla_{\mathbf{k}\perp} E$ might seem rather nebulous;[1] however, it has a simple geometrical interpretation. Consider a second orbit in the same plane as the one of energy E defined at the start of this section, but with energy $E + \delta E$ (see Fig. 8.1). Let the vector $\delta \mathbf{k}$ join a point $\mathbf{k}$ on the orbit of energy E to a point on the orbit of energy $E + \delta E$, and let it also be perpendicular to the orbit of energy E at $\mathbf{k}$. If δE is small, then

$$\delta E = \nabla_{\mathbf{k}} E . \delta \mathbf{k} = \nabla_{\mathbf{k}\perp} E . \delta \mathbf{k}. \qquad (8.5)$$

As $\nabla_{\mathbf{k}}$ is perpendicular to surfaces of constant energy, then $\nabla_{\mathbf{k}\perp}$ is perpendicular to the orbit of energy E and therefore parallel to $\delta \mathbf{k}$. Equation 8.5 becomes

$$\delta E = |\nabla_{\mathbf{k}\perp} E| \delta k, \qquad (8.6)$$

so that eqn 8.4 can be rewritten

$$t_2 - t_1 = \frac{\hbar^2}{eB} \frac{1}{\delta E} \int_{\mathbf{k}_1}^{\mathbf{k}_2} \delta k \, dk. \qquad (8.7)$$

The integral in eqn 8.7 is the area of the k-space plane between orbits E and $E + \delta E$; therefore if $\delta E \to 0$ then

$$t_2 - t_1 = \frac{\hbar^2}{eB} \frac{\partial A_{1,2}}{\partial E}, \qquad (8.8)$$

where $\frac{\partial A_{1,2}}{\partial E}$ is the rate at which the orbit from $\mathbf{k}_1$ to $\mathbf{k}_2$ sweeps out area in k-space as E increases.

In most cases it is going to be useful to work in terms of closed orbits (cyclotron orbits) in k-space, i.e. closed paths where $\mathbf{k}_1 = \mathbf{k}_2$. In this case A becomes the area in k-space of the closed orbit; this will depend on E and the component of $\mathbf{k}$ parallel to $\mathbf{B}$, which we denote $k_{\parallel}$. The period τ_c of the closed orbit is

$$\tau_c = \frac{\hbar^2}{eB} \frac{\partial A(E, k_{\parallel})}{\partial E}; \qquad (8.9)$$

turning this into an angular frequency ω_c (the **cyclotron frequency**), we obtain

$$\omega_c = \frac{eB}{m^*_{\text{CR}}}, \qquad (8.10)$$

where

$$m^*_{\text{CR}} = \frac{\hbar^2}{2\pi} \frac{\partial A(E, k_{\parallel})}{\partial E}. \qquad (8.11)$$

The quantity m^*_{CR} defined in eqn 8.11 is known as the **cyclotron mass**. It is easy to show that for a free-electron system this quantity is just m_e, so that eqn 8.10 yields the classical cyclotron frequency $\omega_c = eB/m_e$.[2] Similarly, in the case of a constant, isotropic effective mass m^*, it can be shown in a straightforward manner that $\omega_c = eB/m^*$.

[1] Or even 'nablaous'.

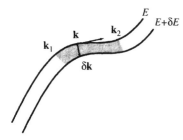

Fig. 8.1 Geometrical interpretation of parameters used in the derivation of the cyclotron frequency; the two curves represent constant energy orbits of energy E and $E + \delta E$ in the plane perpendicular to the magnetic field. The electron traverses from $\mathbf{k}_1$ to $\mathbf{k}_2$ on the orbit of energy E; $\delta \mathbf{k}$ is perpendicular to the orbit of energy E and connects it to the orbit of energy $E + \delta E$. The small arrow perpendicular to $\delta \mathbf{k}$ shows the instantaneous direction of dk which runs along the orbit in the integral of eqn 8.4. The shaded area is the area of the k-space plane between orbits E and $E + \delta E$.

[2] See, for example, *Electricity and magnetism*, by B.I. Bleaney and B. Bleaney, third edition (Oxford University Press, Oxford, 1989) page 126.

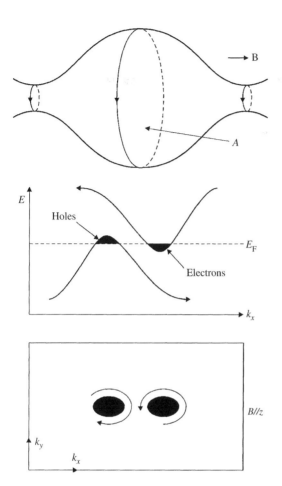

Fig. 8.2 Orbits on a Fermi surface section are in planes perpendicular to **B**. Here, A represents the k-space cross-sectional area of the orbit.

Fig. 8.3 The use of the sign of $(\partial A/\partial E)$ to identify a section of Fermi surface as hole-like or electron-like. The top half of the figure shows a schematic bandstructure with two bands crossing the Fermi energy E_F. E_F crosses the left-hand band near its maximum; the band is almost full, with holes (empty states) at the top of the band surrounded by filled states. E_F crosses the right-hand band near its minimum, so that there are electrons in the bottom of the band surrounded by empty states. The lower half of the figure shows the corresponding Fermi-surface cross-sections. In the left-hand case, as E_F increases, the area of the section of Fermi surface decreases (hole-like). In the right-hand case, as E_F increases, the area of the section of Fermi surface increases (electron-like). The arrows show the opposite senses of cyclotron motion about the Fermi-surface sections.

8.2.3 Orbits on a Fermi surface

Chapters 1–5 showed that the Fermi surface of a metal is a constant–energy surface *par excellence*. Therefore, if the material subjected to the magnetic field is metallic (i.e. possesses sections of Fermi surface), then Section 8.2.1 implies that the electrons will perform orbits in k-space about Fermi-surface cross-sections perpendicular to **B** (see Fig. 8.2). Equations 8.10 and 8.11 can therefore give information about the cross-sectional areas of a section of Fermi surface. Note that the *sign* of $(\partial A/\partial E)$ allows the carriers on a section of Fermi surface to be identified as hole-like $((\partial A/\partial E) < 0)$ or electron-like $((\partial A/\partial E) > 0$ in a very appealing manner (see Fig. 8.3).

8.3 The introduction of quantum mechanics

8.3.1 Landau levels

A general proof of the quantised motion of a carrier in a magnetic field in an arbitrarily-shaped band is difficult. In this section, I shall use a band defined by the effective-mass tensor of Exercise 5.2, Chapter 5, with its minimum centred on $k = 0$ (a good approximation to many of the band extrema in semiconductors) and derive some analytical solutions for the eigenenergies of the electrons (the **Landau levels**). The anisotropic band also yields some useful insights about generalised bandstructure.

For further generality, Section 8.3.2 will derive the Landau quantisation of arbitrarily-shaped bands in the limit that $\hbar\omega_c \ll E_F$.

The Hamiltonian for electrons in the band defined by the effective mass tensor of Exercise 5.2, with its minimum centred on $k = 0$ is

$$\left\{ \frac{\mathcal{P}_x \mathbf{e}_1}{(2m_1)^{\frac{1}{2}}} + \frac{\mathcal{P}_y \mathbf{e}_2}{(2m_2)^{\frac{1}{2}}} + \frac{\mathcal{P}_z \mathbf{e}_3}{(2m_3)^{\frac{1}{2}}} \right\}^2 \psi = E\psi, \qquad (8.12)$$

where the $\mathcal{P}$ are one-dimensional momentum operators; i.e. all of the effects of the crystalline potential have been absorbed into the bandstructure, which is defined by the effective masses. Substitution of a plane wave solution into eqn 8.12 yields energies of the correct form, i.e.

$$E = \frac{\hbar^2 k_x^2}{2m_1} + \frac{\hbar^2 k_y^2}{2m_2} + \frac{\hbar^2 k_z^2}{2m_3}. \qquad (8.13)$$

Consider a magnetic field **B** directed along the z ($\mathbf{e}_3$) axis. In addition, remember that $\mathbf{B} = \nabla \times \mathbf{A}$, where **A** is the magnetic vector potential. One particular **A** which gives the correct **B** is

$$\mathbf{A} = (0, Bx, 0); \qquad (8.14)$$

this choice is known as the **Landau gauge**. For an electron in a magnetic field,

$$\mathcal{P} \to (\mathcal{P} + e\mathbf{A}) \qquad (8.15)$$

(see Appendix G). Making such a substitution into eqn 8.12 gives

$$\left\{ \frac{\mathcal{P}_x \mathbf{e}_1}{(2m_1)^{\frac{1}{2}}} + \frac{(\mathcal{P}_y + eBx)\mathbf{e}_2}{(2m_2)^{\frac{1}{2}}} + \frac{\mathcal{P}_z \mathbf{e}_3}{(2m_3)^{\frac{1}{2}}} \right\}^2 \psi = E\psi. \qquad (8.16)$$

All that we have done is introduce an extra term with x in it; as $[\mathcal{P}_y, x] = [\mathcal{P}_z, x] = 0$,[3] $\mathcal{P}_y$ and $\mathcal{P}_z$ still commute with the Hamiltonian; their associated physical quantities are thus constants of the motion. The operators $\mathcal{P}_y$ and $\mathcal{P}_z$ in eqn 8.16 can therefore be replaced by their constant values $\hbar k_y$ and $\hbar k_z$ respectively. Equation 8.16 becomes

$$\left\{ \frac{\mathcal{P}_x \mathbf{e}_1}{(2m_1)^{\frac{1}{2}}} + \frac{(\hbar k_y + eBx)\mathbf{e}_2}{(2m_2)^{\frac{1}{2}}} + \frac{\hbar k_z \mathbf{e}_3}{(2m_3)^{\frac{1}{2}}} \right\}^2 \psi = E\psi. \qquad (8.17)$$

Squaring the bracket, remembering that the Cartesian unit vectors $\mathbf{e}_j$ are mutually perpendicular (i.e. $\mathbf{e}_j . \mathbf{e}_l = \delta_{jl}$), gives

$$\left\{ \frac{\mathcal{P}_x^2}{2m_1} + \frac{(\hbar k_y + eBx)^2}{2m_2} + \frac{\hbar^2 k_z^2}{2m_3} \right\} \psi = E\psi. \qquad (8.18)$$

Making the substitutions $E' = E - (\hbar^2 k_z^2 / 2m_3)$ and $x_0 = -(\hbar k_y / eB)$, and rearranging yields

$$\left\{ \frac{\mathcal{P}_x^2}{2m_1} + \frac{e^2 B^2}{2m_2}(x - x_0)^2 \right\} \psi = E'\psi. \qquad (8.19)$$

---
[3] These are standard commutation relationships; see e.g. *Quantum mechanics*, by Stephen Gasiorowicz (Wiley, New York, 1974) pages 51, 141.

Equation 8.19 looks exactly like the Hamiltonian of a one-dimensional harmonic oscillator, i.e.,

$$\{\frac{\mathcal{P}_x^2}{2m} + \frac{1}{2}m\omega^2(x-x_0)^2\}\psi = E'\psi, \qquad (8.20)$$

with $\omega = eB/(m_1 m_2)^{\frac{1}{2}}$; thus $E' = (l + \frac{1}{2})\hbar\omega$, with $l = 0, 1, 2, 3.....$ The energy levels of the electron are therefore

$$E(l, B, k_z) = \frac{\hbar^2 k_z^2}{2m_3} + (l + \frac{1}{2})\hbar\omega, \qquad (8.21)$$

with

$$\omega = eB/(m_1 m_2)^{\frac{1}{2}} \equiv \omega_c. \qquad (8.22)$$

The equivalence of ω and ω_c, the cyclotron frequency for this geometry (i.e. $\omega_c = eB/m^*_{CR}$, with $m^*_{CR} = \frac{\hbar^2}{2\pi}\frac{\partial A(E, k_\parallel)}{\partial E}$), is proved in Exercise 8.3.

The following points may be deduced from eqn 8.21.

- The energy of the electron's motion in the plane perpendicular to **B** is completely quantised. These quantised levels are known as *Landau levels*.
- The k-space areas of the orbits in the plane perpendicular to **B** are also quantised (this is easy to work out in the present case and is left as an exercise); thus, allowed orbits fall on 'Landau tubes' in k-space with quantised cross-sectional area.
- The energy 'quantum' for the in-plane motion appears to be $\hbar \times$(the semiclassical cyclotron frequency).
- The motion parallel to **B** is unaffected (cf. Section 8.2.1).

We shall consider the second point in more detail (and more generally) in the following section.

8.3.2 Application of Bohr's correspondence principle to arbitrarily-shaped Fermi surfaces in a magnetic field

In Section 8.4 below we are going to consider the effect of the quantisation of the k-space orbits caused by **B** on the Fermi surfaces of metals; i.e. we shall be dealing with Landau levels which cut the Fermi surface. The Fermi surfaces of real metals are in general not as simple as the case dealt with in the previous section. Therefore, in order to treat an arbitrily-shaped band, resulting in an arbitrarily-shaped Fermi surface, we shall use Bohr's correspondence principle, which states in this context that the difference in energy of two adjacent levels is $\hbar$ times the angular frequency of classical motion at the energy of the levels.

The correspondence principle is only valid for levels with very large quantum numbers. However, this is not really a problem, as we are going to be dealing with Landau levels with energies comparable to E_F; in most metals E_F is $\sim$several eV, whereas the cyclotron energy of free electrons is $\hbar eB/m_e \approx 1.16 \times 10^{-4}$ eV at $B = 1$ T. Laboratory fields are usually ~ 10 T, and so the quantum numbers of Landau levels with energies $\sim E_F$ will be $l \sim 10^4$; the correspondence principle should work reasonably well.

The classical frequency at the energy of interest (i.e. E_F) is given by eqn 8.10

$$\omega_c = \frac{eB}{m^*_{CR}}, \qquad (8.23)$$

with m^*_{CR} given by eqn 8.11 as follows

$$m^*_{CR} = \frac{\hbar^2}{2\pi} \frac{\partial A(E, k_{||})}{\partial E}. \qquad (8.24)$$

Here, A is now specifically *the cross-sectional area of the Fermi surface in a plane perpendicular to* **B**. Thus, the separation in energy of two Landau tubes with energies close to E_F is

$$E(l+1, B, k_{||}) - E(l, B, k_{||}) = \frac{\hbar eB}{m^*_{CR}}, \qquad (8.25)$$

with m^*_{CR} defined by eqn 8.24. (It is interesting to compare this equation with eqn 8.21.)

8.3.3 Quantisation of the orbit area

Equation 8.25 may be rewritten as

$$(E(l+1, B, k_{||}) - E(l, B, k_{||}))\frac{\partial A(E)}{\partial E} = \frac{2\pi eB}{\hbar}. \qquad (8.26)$$

Now as the E in $A(E)$ is of order E_F, then the difference between adjacent levels will be $\ll$ the energies of the levels themselves. Thus the approximation

$$\frac{\partial A(E)}{\partial E} = \frac{A(E(l+1, B, k_{||})) - A(E(l, B, k_{||}))}{E(l+1, B, k_{||}) - E(l, B, k_{||})} \qquad (8.27)$$

may be made, so that eqn 8.26 becomes

$$A(E(l+1, B, k_{||})) - A(E(l, B, k_{||})) = \frac{2\pi eB}{\hbar}. \qquad (8.28)$$

This states that classical orbits at adjacent allowed energies and the same $k_{||}$ enclose k-space areas that differ by a fixed amount δA, with

$$\delta A = \frac{2\pi eB}{\hbar}. \qquad (8.29)$$

Putting this another way, the k-space area enclosed by an orbit of allowed energy and $k_{||}$ must be given by

$$A(E(l+1, B, k_{||})) = (l+\lambda)\delta A = (l+\lambda)\frac{2\pi eB}{\hbar} \qquad (8.30)$$

for large l, where λ is a constant ~ 1.

As in Section 8.3.1, we have found 'Landau tubes' which define the allowed k-space orbits; rather than being quasicontinuous, the Fermi surface will now be composed of occupied states lying on Landau tubes (see Fig. 8.4). Note that the k-space area of each tube increases with B; as we shall see in the sections below, this means that tubes 'pop out' of the Fermi surface as B increases.

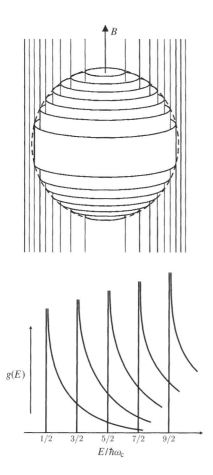

Fig. 8.4 Schematic of a Fermi sphere rearranged into Landau tubes.

Fig. 8.5 Schematic of the electronic density of states in a magnetic field. The 'sawtooth' structure results from a one-dimensional density of states (from the motion parallel to **B**) superimposed on the δ-function-like Landau levels.

8.3.4 The electronic density of states in a magnetic field

The above sections have shown that the energy of the band electron is completely quantised into Landau levels in the plane perpendicular to **B**, whilst the motion parallel to **B** remains unconstrained. This means that the density of states of the electrons is an infinite 'ladder' of Landau levels, each with a one-dimensional density of states function, due to the motion parallel to **B**, superimposed.[4] The overall density of states will therefore be of the form $g(E) = \sum_{l=0}^{\infty} C(E - E(l, B))^{-\frac{1}{2}}$, where $E(l, B)$ is the energy of the Landau level (i.e. the lowest energy point on the Landau tube) and C is a constant; it is shown schematically in Fig. 8.5.

[4] In other words, as far as the density of states is concerned, the magnetic field reduces the effective dimensionality of the electron system by 2.

8.4 Quantum oscillatory phenomena

Each time one of the sharp peaks in the electronic density of states moves through the chemical potential μ,[5] there will be a modulation of the density of states at μ. Since almost all of a metal's properties depend on the density of states at μ, we expect this to affect the behaviour of the metal in some way.

Figure 8.6 shows a Landau tube superimposed on constant energy surfaces E and $E + \delta E$. The left-hand part of the figure shows the general case; as

[5] Remember from earlier chapters that $E_F \equiv \mu(T = 0)$. In many books, quantum oscillations are said to result from 'Landau levels passing through the Fermi energy'. This is only strictly true at $T = 0$; the chemical potential is the important energy as far as the electrons are concerned. In metals at low temperatures there will be a very small difference between E_F and μ; nevertheless, one should not tolerate sloppy terminology.

Fig. 8.6 Schematic of Landau tubes crossing surfaces of constant energy E and $E + \delta E$. (a) shows the Landau tube on its own; (b) shows the Landau tube intersecting constant energy surfaces E and $E + \delta E$ when the tube does not correspond to an extremal orbit; (c) shows the Landau tube intersecting constant energy surfaces E and $E + \delta E$ when the tube corresponds to an extremal orbit. In (c), a much greater area of the cylinder (and therefore a larger number of states) lies between E and $E + \delta E$ than is the case in (b).

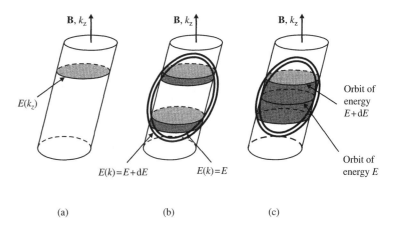

the field increases, the tube sweeps outwards, so that some of its states are always crossing the constant energy surfaces. However, in the right-hand part of the figure, the tube reaches an **extremal cross-section** of the constant energy surface E. In this case, a much larger area of the cylinder (and therefore a much greater number of states, which lie uniformly in k-space) lies between E and $E + \delta E$. Returning to the case of the Fermi surface, we can therefore say that the maximum effect of the Landau quantisation will occur when a Landau tube (and its associated peak in the density of states) crosses an extremal cross-section of a Fermi surface. By analogy with the discussion of classical orbits about the Fermi surface in Section 8.2.1, such cross-sections are often known as **extremal orbits**.

The Landau tube crosses an extremal orbit/cross-section when

$$(l + \lambda)\frac{2\pi e B}{\hbar} = A_{\text{ext}}, \tag{8.31}$$

where A_{ext} is the k-space area of the extremal cross-section of the Fermi surface in a plane perpendicular to **B**. Thus, the metal's properties will oscillate as **B** changes, with a period given by

$$\Delta\left(\frac{1}{B}\right) = \frac{2\pi e}{\hbar}\frac{1}{A_{\text{ext}}}. \tag{8.32}$$

Such oscillations are known generically as **magnetic quantum oscillations**.

Often it is more convenient to describe a series of quantum oscillations using a frequency F, also known as the **fundamental field**, B_{F};

$$F \equiv B_{\text{F}} = \frac{1}{\Delta(1/B)} = \frac{\hbar}{2\pi e}A_{\text{ext}}. \tag{8.33}$$

Note that

- the fundamental field of the quantum oscillations is determined solely by the Fermi surface extremal area and fundamental constants;
- several different frequencies (fundamental fields) of quantum oscillation may be simultaneously present for a given orientation of **B**, corresponding to different possible extremal orbits (see Fig. 8.7); a famous example is the simultaneous observation of frequencies from 'neck and belly' orbits in Cu and Au;

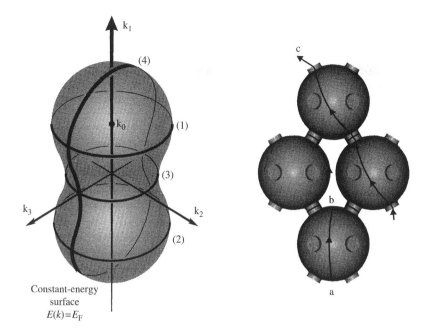

Fig. 8.7 Illustration of the types of extremal orbit about Fermi surfaces which can give rise to quantum oscillations. The left-hand figure shows that for **B** parallel to k_1 there will be three extremal orbits ((1), (2) and (3)); as (1) and (2) have the same area, there will be two series of quantum oscillation (i.e. two different frequencies will be present). However, for **B** parallel to k_2, there is only one extremal orbit (4); a similar situation would apply if **B** were parallel to k_3. The right-hand figure helps to visualise the different types of orbit possible in metals possessing necks and bellies in their Fermi surfaces, such as copper, silver or gold. The orbit labelled a is an orbit around the 'belly' of the Fermi surface. If the magnetic field is applied out of the page, then an extremal 'hole-like' orbit (labelled b) will occur around the bone-shaped space in the middle; this is known as the **dog's bone** orbit! In orbit c, electrons run across the bellies and necks in an extended or **open** orbit; *this does not give rise to quantum oscillations*, as there is no closed orbit with which a frequency can be associated. If the magnetic field is applied along one of the neck directions, extremal orbits about both neck and belly are possible, leading to two frequencies of quantum oscillations. Note that different field orientations are required to observe the different cases; the orbits in k-space are *always* in planes perpendicular to **B**.

- a measurement of the observed frequencies as a function of magnetic field orientation allows the Fermi-surface shape to be mapped out;
- open orbits do not give rise to quantum oscillations.[6]

[6] However, open orbits do lead to a very interesting quantum phenomenon which has recently been observed in high-frequency experiments; see A. Ardavan *et al.*, *Phys. Rev. B* **60**, 15500 (1999); *Phys. Rev. Lett.* **81**, 713 (1998).

8.4.1 Types of quantum oscillation

As the electronic density of states at E_F determines most of a metal's properties, virtually all properties will exhibit quantum oscillations in a magnetic field. Examples include[7]

- oscillations of the magnetisation (the de Haas–van Alphen effect);
- oscillations of the magnetoresistance (the Shubnikov–de Haas effect);
- oscillations of the sample length;
- oscillations of the sample temperature;

[7] Some pictures of typical data are shown in *Solid state physics*, by N.W Ashcroft and N.D. Mermin (Holt, Rinehart and Winston, New York, 1976) pages 266–268.

Fig. 8.8 Coils for the observation of the de Haas–van Alphen effect. For each set, c is the coil flux-linked to the sample and c^0 is the compensating coil, only weakly flux-linked to the sample: (a) c is wound around the centre of a disc-like sample; (b) c is wound around a sample cut into a cylinder; (c) a small sample inside colinear coils c and c^0; (d) small sample inside coaxial coils c and c^0; (e) side and top views of a coil evaporated down onto a thin sample; (f) as (e) but with only a single annular turn. In both (e) and (f) c^0 has not been shown. (Based on figures in *Magnetic oscillations in metals*, by David Shoenberg (Cambridge University Press, Cambridge, 1984).)

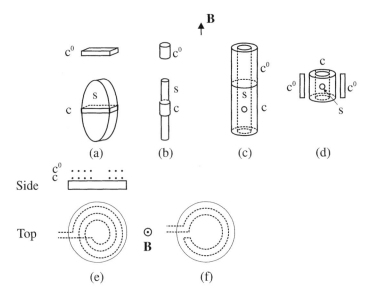

- oscillations in the ultrasonic attenuation;
- oscillations in the Peltier effect and thermoelectric voltage;
- oscillations in the thermal conductivity.

8.4.2 The de Haas–van Alphen effect

The de Haas–van Alphen effect is perhaps the most significant of the quantum oscillatory phenomena in metals; as the magnetisation M is a thermodynamic function of state;[8] it can be directly related to the density of states and Fermi–Dirac distribution function of the electrons in a metal, *without additional assumptions*. This means that theoretical models for the Fermi surface can be checked in a very rigorous manner.[9]

Figure 8.8 shows a schematic of the types of coil most commonly used to observe the de Haas–van Alphen effect. Each system consists of a coil c which contains the sample and a *compensating coil* c^0; both are placed in a time-varying magnetic field. The coils c and c^0 are connected in series-opposition and carefully balanced so that in the absence of a sample the voltage induced by ${\rm d}B/{\rm d}t$ in c^0 exactly cancels that in c. When the sample is placed in c, then c and c^0 are no longer balanced; a voltage V is induced, where

$$V = \alpha \frac{{\rm d}M}{{\rm d}B}\frac{{\rm d}B}{{\rm d}t}. \tag{8.34}$$

Here α is a parameter depending on the geometry of the coil and sample. The de Haas–van Alphen oscillations occur in ${\rm d}M/{\rm d}B$ and thus will be visible in V.[10]

The field is provided in two ways.

- For fields of up to ~ 20 T, superconducting magnets are generally used. The magnet consists of two parts, a large outer coil providing a static field and an inner coil which provides a small modulation field (i.e. ${\rm d}B/{\rm d}t$) of mT amplitude at a few tens of Hz.

[8] See e.g. *Equilibrium thermodynamics* by C.J Adkins, third edition (Cambridge University Press, Cambridge 1983) page 6 and Appendix.

[9] For enthusiasts, a relatively recent example of this is given in 'A numerical model of quantum oscillations in quasi-two-dimensional organic metals in high magnetic fields', by N. Harrison *et al.*, *Physical Review B* **54**, 9977 (1996).

[10] de Haas–van Alphen oscillations are also sometimes observed using a *torque magnetometer*; see J.S. Brooks *et al.*, *Rev. Sci. Instrum.* **64**, 3248 (1987).

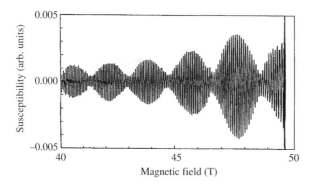

Fig. 8.9 de Haas–van Alphen oscillations in Pt at 4.2 K with **B** parallel to [111]. The data have been recorded using a pulsed magnetic field. Note the presence of two frequencies of oscillations due to two extremal orbits about the Fermi surface. The y axis, labelled 'susceptibility', represents the voltage V induced in the coil divided by (dB/dt), leaving a quantity proportional to the differential susceptibility (dM/dB) (see eqn 8.34). (Data from S. Askenazy, *Physica B* **201**, 26 (1994).)

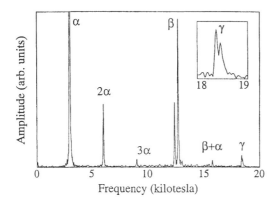

Fig. 8.10 Fourier transform of de Haas–van Alphen oscillations of Sr_2RuO_4. Fourier transformation is routinely used to help understand de Haas–van Alphen oscillations. The oscillation data (recorded digitally) are processed so that they are equally spaced in $1/B$; the resulting file is Fourier transformed using a desk-top computer. Each extremal cross-section of the Fermi-surface generates a peak in the Fourier transform, the frequency of which is proportional to the k-space area. Oscillations which are non-sinusoidal in shape generate harmonics in the Fourier transform; frequency mixing effects also occur, resulting in sum frequencies. Three Fermi-surface cross-sections (labelled α, β and γ) are shown, along with some harmonics of α and a sum frequency $\beta + \alpha$. (Data from A.P. McKenzie *et al.*, *J. Phys. Soc. Jpn.* **67**, 385 (1998)).

- Pulsed magnets are used for fields of up to ~ 60 T; in this case the magnetic field rises to its maximum value in a few milliseconds, so that dB/dt is large (there is no need for an additional modulation coil).

Some typical de Haas–van Alphen data are shown in Figs. 8.9 and 8.10.

The coils are often mounted on a rotation stage so that the sample can be rotated *in situ*, allowing measurements to be made with **B** applied at several orientations to the crystal axes. The coils and sample are mostly mounted in ^{3}He cryostats or dilution refrigerators, as mK temperatures are needed to observe the oscillations in many cases (see below).

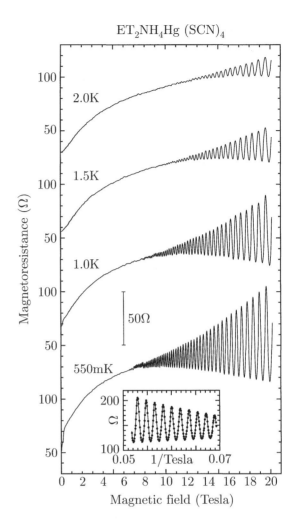

Fig. 8.11 Shubnikov–de Haas oscillations in the organic molecular solid α-(BEDT-TTF)$_2$NH$_4$Hg(SCN)$_4$; data for different temperatures have been offset for clarity. The inset shows that the oscillations are periodic in $1/B$. (Data from J. Singleton, *Reports on Progress in Physics* **63**, 1111 (2000).)

8.4.3 Other parameters which can be deduced from quantum oscillations

So far we have concentrated on the information about the Fermi-surface shape provided by the quantum oscillations. However, other valuable information can be deduced. Figure 8.11 shows Shubnikov–de Haas oscillations in an organic solid (see Chapter 7 for more details of this sort of conductor). Two things are immediately obvious.

(1) **The oscillations grow in amplitude as the temperature is lowered.** This occurs because finite temperatures 'smear out' the edge of the Fermi–Dirac distribution function which separates filled and empty states (see Fig. 1.7). At $T = 0$ the Landau tubes will pop out of a very sharp Fermi surface; at finite T the surface will be 'fuzzy'. The modulation of the density of states around μ caused by the Landau tubes will therefore be less significant. The thermal smearing implies that $\hbar\omega_c$ must be greater than $\sim k_B T$ for oscillations to be observed, i.e. low temperatures are needed.

As the decline in intensity is caused by the Fermi–Dirac distribution function, it can be modelled rather well, and an analytical expression for the temperature dependence of the oscillation amplitude, valid as long as the oscillation amplitude is small, has been derived[11]

$$\text{Amplitude} \propto \frac{\chi}{\sinh \chi}, \qquad (8.35)$$

where $\chi = 14.7 m^*_{\text{CR}} T/B$ (with T in Kelvin and B in Tesla). It is fairly obvious that the parameter χ is proportional to the ratio $k_B T / \hbar \omega_c$; i.e. what we are doing is comparing $k_B T$ and $\hbar \omega_c$. Fits of the oscillation amplitude to eqn 8.35 can therefore be used to give m^*_{CR}.

[11] See *Magnetic oscillations in metals*, by David Shoenberg (Cambridge University Press, Cambridge, 1984) Chapter 2.

(2) **The oscillations grow in amplitude as B increases.** This occurs because scattering causes the Landau levels to have a finite energy width $\sim \hbar/\tau$. As B increases, $\hbar \omega_c$ will increase, so that the broadened Landau levels will become better resolved; theoretical studies have shown that this causes the oscillation amplitude to vary with B as follows (again, the equation is only valid for small oscillations)

$$\text{Amplitude} \propto B^{-r} \, e^{-\frac{\pi}{\omega_c \tau}}, \qquad (8.36)$$

where r is a number ~1. Therefore the scattering rate τ^{-1} can be extracted.

In addition, information about the g-factor of the electrons can be obtained from the oscillations under certain circumstances.[12]

[12] See *Magnetic oscillations in metals*, by David Shoenberg (Cambridge University Press, Cambridge, 1984) and J. Singleton, *Reports on progress in physics* **63**, 1111 (2000).

8.4.4 Magnetic breakdown

The frequencies observed in the quantum oscillations correspond to the areas of closed semiclassical orbits about the extremal areas of the Fermi surface. However, under certain circumstances, much higher frequency oscillations can become apparent at high fields, corresponding to larger k-space areas. These are caused by **magnetic breakdown**.

Figure 8.12 shows a simple Fermi surface cross-section with two open semiclassical orbits (incapable of giving quantum oscillations) and a closed orbit (capable of giving quantum oscillations); the arrows show the electron trajectories. At low fields, one frequency of oscillations corresponding to the closed orbit will be seen. At high fields, the electrons have sufficient cyclotron energy to tunnel *in k-space* from one part of the Fermi surface to another. Therefore they can now describe much larger closed k-space orbits, leading to higher-frequency quantum oscillations. In effect, the magnetic field is beginning to break down the arrangement of the bandstructure at zero field.

We shall see an example of magnetic breakdown later in this chapter.

8.5 Cyclotron resonance

It is possible to make a direct measurement of ω_c using millimetre-waves or far-infrared radiation to excite transitions between Landau levels. Such an experiment is known as **cyclotron resonance**. The conditions for observing cyclotron resonance in metals and semiconductors are rather different, and the two cases are described below.

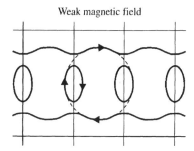

Weak magnetic field

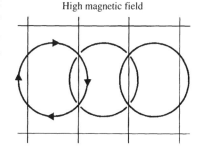

High magnetic field

Fig. 8.12 Explanation of magnetic breakdown. Above: semiclassical orbits on a simple Fermi surface at low fields; below: magnetic breakdown at high fields. The magnetic field is perpendicular to the plane of the page.

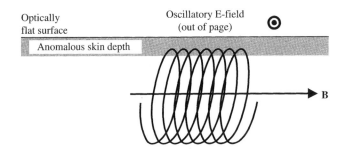

Fig. 8.13 Geometry of a cyclotron resonance experiment in a metal. The shaded region indicates the skin depth for the radiation; the helical orbits of the electrons are shown schematically.

8.5.1 Cyclotron resonance in metals

The field components of electromagnetic radiation decay with distance z into a conducting material as $\exp(-z/\delta)$, where $\delta = (\frac{1}{2}\sigma\omega\mu_{\rm r}\mu_0)^{-\frac{1}{2}}$ is the *skin depth* or anomalous skin depth; here σ is the conductivity of the material, $\mu_{\rm r}\mu_0$ is its permeability and ω is the angular frequency of the radiation. Metals have rather high conductivities and hence small skin depths; typical values for copper are $\delta \approx 7 \times 10^{-5}$m at $\omega/2\pi \approx 1$ MHz and $\delta \approx 7 \times 10^{-7}$m at $\omega/2\pi \approx 10$ GHz.[13]

Typical effective masses in metals combined with the magnetic fields readily available in laboratories have meant that frequencies $\omega/2\pi \sim 1 - 100$ GHz have tended to be applied in cyclotron resonance experiments (see Section 8.8). At such frequencies, radiation cannot penetrate far into the crystal, dictating the geometry of the cyclotron resonance measurement (see Fig. 8.13). The magnetic field is applied parallel to a surface of the crystal, which is placed in a region of oscillating electric field in a resonant cavity; **E** of the radiation is arranged to be perpendicular to **B** and parallel to the surface. If the frequency of the radiation $\omega = j\omega_{\rm c}$, where j is an integer, then the electrons will receive a 'kick' from the radiation's electric field every time they come within a skin depth of the surface; this results in absorption of energy. Usually ω is kept constant and the field is swept, so that absorptions are uniformly spaced in $1/B$. As in the case of the quantum oscillations, low temperatures are required. A magnetic field which is very accurately parallel to the sample's surface is required for a successful measurement. This type of experiment is often referred to as the Azbel'–Kaner geometry.

[13] The skin depth is derived in *Electricity and magnetism*, by B.I. Bleaney and B. Bleaney, third edition (Oxford University Press, Oxford, 1989) page 236.

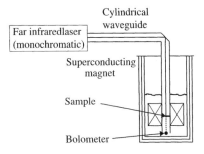

Fig. 8.14 Schematic of a cyclotron resonance experiment in a semiconductor.

8.5.2 Cyclotron resonance in semiconductors

The number densities of carriers in semiconductor samples are much lower than those in metals, so that the radiation can completely penetrate even large samples. A simple transmission arrangement is usually adopted (see Fig. 8.14).

The cyclotron resonance is usually recorded by measuring the sample transmission; experiments are carried out either by fixing the magnetic field and varying the energy of the radiation (see Figure 8.15), or by using a fixed-frequency source (such as a far-infrared laser, which can give a number of monchromatic laser lines in the wavelength range 40 μm to 1000 μm) and sweeping the magnetic field (see Figs. 8.14 and 8.16).[14] The magnetic field is usually provided by a superconducting magnet ($B = 0- \sim 20$ T). As the whole of the cyclotron orbit experiences the electric field of the radiation (cf. the situation in metals; see above), the quantum-mechanical selection rule

[14] An alternative arrangement for semiconductor cyclotron resonance (rather an old-fashioned one) is given in *Electricity and magnetism*, by B.I. Bleaney and B. Bleaney, third edition (Oxford University Press, Oxford, 1989) pages 729–734; the description also provides a more rigorous derivation of the linewidth of the cyclotron resonance than the one given here.

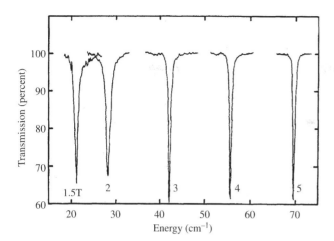

Fig. 8.15 Cyclotron resonances in a GaAs-(Ga,Al)As heterojunction at $T = 4.2$ K; the magnetic field has been applied perpendicular to the two-dimensional electron layer, which has an areal carrier density of $\sim 9 \times 10^{10}$cm^{-2}. The data have been recorded by fixing the magnetic field at 1.5, 2, 3, 4, and 5 T; at each field, a Fourier-transform spectrometer has then been used to record the transmission of the sample as a function of energy. (Data recorded by G. Wiggins, University of Oxford. For further data of this type, see e.g. C.J.G.M. Langerak et al., Phys. Rev. B. **38**, 13133 (1988).)

for the Landau-level quantum number ($\Delta l = \pm 1$) holds. Thus the resonance condition is

$$\omega = \omega_c = \frac{eB}{m^*_{CR}}, \quad (8.37)$$

where ω is the angular frequency of the radiation.

The linewidth of the resonance can give information about the scattering rate τ^{-1}. The scattering induces a frequency uncertainty $\Delta\omega_c \sim \tau^{-1}$. If the experiment is a fixed-frequency, swept-field one, this translates to an uncertainty (i.e. resonance width) in magnetic field of $\Delta B = \Delta\omega_c m^*_{CR}/e \sim m^*_{CR}/e\tau$.[15]

In the case of degenerate semiconductor systems (e.g. heterojunctions), free carriers are present even at low temperatures; typical cyclotron resonance data are shown in Fig. 8.15. However, in lightly doped samples, the carriers must be excited into the bands by either raising the temperature to cause the impurities to ionise (but not so far as to broaden the Landau levels) or by additionally illuminating the sample with above-band-gap radiation.

Figure 8.16 shows an example of the latter technique applied to a single crystal of Ge. In addition to separate resonances caused by light and heavy holes, three electron cyclotron resonances are observed. This occurs becaue the anisotropic conduction-band minima in Ge lie along the [111] axes (see Fig. 6.5); in Fig. 8.16 the static magnetic field makes three different angles with the four [111] axes. The conduction-band minima in Ge can be described using an effective mass tensor (see Chapter 5); if one of the principal axcs of the coordinate system is directed along a [111] axis, then the tensor is diagonal. Under such circumstances, eqn 5.28 shows that the cyclotron resonance occurs at $\omega = \frac{eB}{m^*}$, with

$$m^* = \left(\frac{B^2 m_1 m_2 m_3}{m_1 B_x^2 + m_2 B_y^2 + m_3 B_z^2} \right)^{\frac{1}{2}},$$

where the components of the magentic field (B_1, B_2, B_3) are defined with respect to the principal axes which make the tensor diagonal. Thus, the cyclotron resonance frequency depends on the orientation of the static magnetic field

[15] Provided that $m^*_{CR} \approx m^*$, the carrier effective mass for linear motion (and this is often true in direct-gap semiconductors like GaAs) it is apparent that $1/\Delta B \sim$ the carrier mobility (see eqn 9.12).

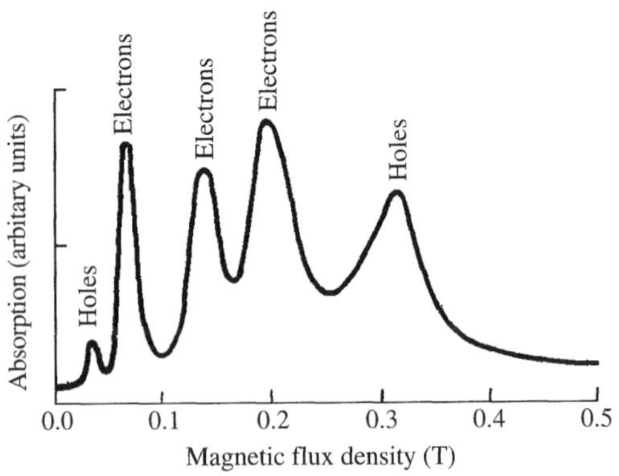

Fig. 8.16 Absorption caused by cyclotron resonance of electrons and holes in a single crystal of Ge; the electrons and holes are present because the sample has been illuminated by above-band-gap light. The microwave radiation frequency has been fixed at 24 GHz and the magnetic field, applied in a (110) plane at 60° to a [100] axis, has been swept. Cyclotron resonances due to light (low field) and heavy (high field) holes are visible, as are three electron cyclotron resonances. (Data from Dresselhaus et al., Phys. Rev. **98**, 368 (1955).)

Table 8.1 Effective masses (in units m_e) in Ge and Si determined using cyclotron resonance. The masses m_1, m_2 and m_3 refer to the components of the effective mass tensors for the conduction band; m_{lCR} and m^*_{hCR} are the cyclotron masses of light and heavy holes respectively.

Material	$m_1 = m_2$	m_3	m_{lCR}	m^*_{hCR}
Ge	0.082	1.64	0.044	0.3
Si	0.19	0.98	0.16	0.3

with respect to the crystal axes. Such angle-dependent cyclotron resonance experiments can be used to determine components of the effective mass tensor.

In the case of Ge and Si, it is found that two of the principal values of the effective mass tensor are equal, so that the constant energy surfaces are ellipsoids with circular cross-sections (see Fig. 6.5). The effective masses of Si and Ge determined by cyclotron resonance are shown in Table 8.1.

8.6 Interband magneto-optics in semiconductors

The application of a magnetic field to a semiconductor splits both the valence and conduction bands into Landau levels, and optical transitions can be induced between hole and electron Landau levels. In reasonably simple systems, the selection rule for the Landau level index is $\Delta l = 0$; more subtle selection rules apply in materials with complex bandstructures.[16] Taking the former, simpler case, the transitions therefore have the energies

$$E(l, B) = E_g + \frac{\hbar e B}{m^*_{\text{hCR}}}(l + \frac{1}{2}) + \frac{\hbar e B}{m^*_{\text{cCR}}}(l + \frac{1}{2}), \tag{8.38}$$

where $l = 0, 1, 2, 3..$, and m^*_{cCR} and m^*_{hCR} are the cyclotron masses of the electron and the hole respectively (usually the heavy holes dominate, because of their large density of states). Such a case is dealt with in Exercise 8.1.

[16] This is a consequence of the angular momentum carried by the photon. The valence bands of many semiconductors are based on atomic orbitals with total angular momentum quantum number $J = 3/2$, whereas the conduction band is based on orbitals with $J = 1/2$. Thus the one unit of angular momentum carried by the photon is required to promote an electron across the band gap, leaving nothing to spare to change the Landau-level quantum number l. By contrast, in intraband magneto-optical transitions (cyclotron resonance), the photon's angular momentum is used to change l by one ($\Delta l = \pm 1$).

8.6 *Interband magneto-optics in semiconductors* 101

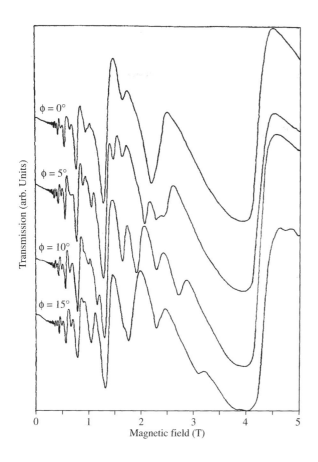

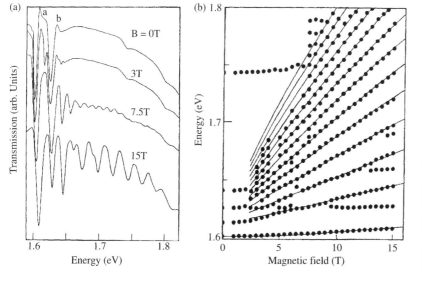

Fig. 8.17 Transmission of PbTe at a fixed photon energy of $h\nu = 0.2077$ eV as a function of magnetic field ($T = 1.9$ K). Data for several orientations of the sample in the external field, which is at all times perpendcular to [111], are shown; ϕ is the angle between the magnetic field and [$\bar{1}10$]. (Data from J. Singleton *et al., J. Phys. C: Solid State Phys.* **19**, 77 (1986).)

Fig. 8.18 (a) Transmission of a 55 ÅGaAs-(Ga,Al)As quantum well at various fixed magnetic fields ($T = 55$ K). The magnetic field has been applied perpendicular to the plane of the quantum well. (b) Resulting fan diagram of transition energies versus magnetic field; points are data, and the lines are a theoretical fit. (From D.C. Rogers *et al., Phys. Rev. B* **34**, 4002–4009 (1986).)

Illustrative experimental data are shown in Figs. 8.17 and 8.18. In Fig. 8.17, the transmission of a thin film of PbTe has been measured using fixed-frequency radiation of energy $h\nu = 0.2077$ eV as the magnetic field is swept. Transitions between valence-band Landau levels and conduction-band Landau levels are observed as dips in the transmission; note that the dips get further apart and better resolved as the field increases. Equation 8.38 gives a qualitative understanding of these observations; as the field increases, the inter-Landau-level transitions will successively pass through the radiation energy $h\nu$, resulting in a spectrum which looks roughly periodic in $1/B$. PbTe has complex, anisotropic conduction and valence-band extrema, so that the conduction and valence-band Landau level energies vary as the orientation of the magnetic field changes; this is illustrated by the rapid change in the spectra as the sample is tilted in the field, showing how a comprehensive study of this kind can map out the band shapes.

In systems which involve excitonic effects, there are usually complications to be taken into account. Figure 8.18(a) shows the transmission of a GaAs-(Ga,Al)As quantum well at four fixed fields as a function of energy. At zero field, only the excitonic transitions from the hole subbands to the conduction-band subbands are observed, but as the field increases, transitions between the heavy-hole Landau levels and the conduction-band Landau levels become visible. Figure 8.18 shows a 'fan diagram' of the transition energies versus magnetic field for the same sample. Note the following:

- the Landau-level transitions no longer extrapolate back to a point (cf. eqn 8.38), but instead evolve directly from the various states of the exciton (this is particularly notable for the lowest ($l = 0$ and $l = 1$) transitions); there is a one-to-one correspondence between the hydrogen-like levels of the exciton and the Landau level transitions (1s evolves into $l = 0$ transition, 2s evolves into $l = 1$ transition etc.;

- the field dependences of the transition energies deviate from straight lines at high energies, curving downwards; this is due to the **nonparabolicity** of the bands as one moves away from the band extrema.

Data such as these can be used to extract the effective masses of the electron and heavy-hole subbands, parameters which describe the nonparabolicity and the exciton binding energy.

A review of recent work in this field is listed under Further Reading.

8.7 Other techniques

The experimental techniques mentioned thus far tend to give information about very restricted regions of the bands concerned. In the case of the magnetic quantum oscillations, electrons very close to the Fermi surface are studied; similarly, the magneto-optical measurements of the semiconductors tend to measure the bands close to their extrema. Furthermore, the methods described above are only usually possible at low temperatures. In the following sections, I shall briefly describe experimental methods which attempt to measure the bandstructure away from these points and which may be used at higher temperatures.

8.7.1 Angle-resolved photoelectron spectroscopy (ARPES)

Angle-resolved photoelectron spectroscopy (ARPES) uses photons with typical energies $h\nu$ in the range ~ 5 eV (ultraviolet) to ~ 1 keV (soft X-ray) to promote electrons from the occupied bands of a solid to unoccupied energy levels from which they can escape. Figure 8.19(a) shows this process schematically. An electron, initially in a state with energy E_i, is excited to an empty state with energy E_f, which is above the vacuum level, E_{vac}; the electron is then able to escape the solid with kinetic energy

$$E_{kin} = E_f - E_{vac} = E_i + h\nu - E_{vac} \tag{8.39}$$

(this is basically the famous Einstein equation describing the photo-electric effect). By measuring E_{kin} and $h\nu$, one can then deduce E_i.

For many years, angle-integrated photoemission, in which all of the electrons of a certain energy are collected irrespective of angle of emission, was used as a way of characterising the bandstructure of solids. In such an experiment, the number of electrons emitted is basically determined by the density of states $g(E)$ at E_i. However, it was always realised that the wavevector of the emitted electron also contained information about the initial state. With the advent of high-intensity synchrotron radiation sources in the late 1970s and early 1980s, the numbers of emitted electrons became sufficient for angle-resolved collection of electrons to be feasible.

In free space, a single electron cannot absorb a photon, since momentum cannot be conserved. However, in a solid, the extra momentum is provided by the surrounding lattice. It turns out[17] that the component of $\mathbf{k}$ parallel to the surface of the crystal is conserved according to the equation

$$\mathbf{k}_i^{\|} = \mathbf{k}_f^{\|} + \mathbf{G}^{\|}, \tag{8.40}$$

where $\mathbf{k}_i$ is the value of $\mathbf{k}$ for the initial state of energy E_i, $\mathbf{k}_f$ is the final value of $\mathbf{k}$ outside the solid and the suffix $\|$ indicates the components parallel to the surface (see Fig. 8.19(b)). Ultraviolet photons have negligible momentum compared to band electrons, so that the photon momentum is neglected in this equation; the lack of periodicity in the direction perpendicular to the surface means that the momentum transfer in this direction is unconstrained. Note that $\mathbf{G}^{\|}$ is a *surface* reciprocal lattice vector, that is, one defined by the symmetry of the crystal at the surface. In many cases, the surfaces of crystals are *reconstructed*, and so $\mathbf{G}^{\|}$ may be rather different from that of the bulk reciprocal lattice, creating a possible source of error in the interpretation of ARPES. As the energy E_{kin} of the emitted electrons is measured, $\mathbf{k}_f^{\|}$ may be deduced using

$$|\mathbf{k}_f^{\|}| = \frac{(2m_e E_{kin})^{\frac{1}{2}}}{\hbar} \sin\theta. \tag{8.41}$$

Equations 8.39, 8.40 and 8.41 are sufficient to reconstruct E_i and $\mathbf{k}_i$ of the initial state.

ARPES has been used extensively with layered compounds in which the bandstructure can be regarded as quasi-two-dimensional; in such materials the deduction of the underlying bandstructure from the emitted electron energies is often relatively simple. Figure 8.20 shows ARPES data from the layered

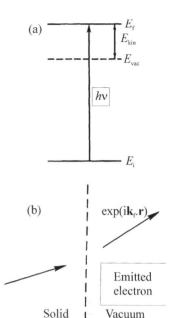

Fig. 8.19 (a) Schematic of ARPES. An electron, initially in a state with energy E_i, is excited by a photon of energy $h\nu$ to an empty state with energy E_f, above the vacuum level, E_{vac}; the electron is then able to escape the solid with kinetic energy $E_{kin} = E_f - E_{vac}$ (b) Simplified geometry of emission process at the surface. The electron is emitted with an energy E_{kin} at an angle θ to the sample surface normal; the component of $\mathbf{k}$ parallel to the surface is conserved according to eqn 8.40.

[17] See F.J. Himpsel, *Advances in Physics* **32**, page 11 (1983) for a simple derivation of the conservation laws.

104 *Measurement of bandstructure*

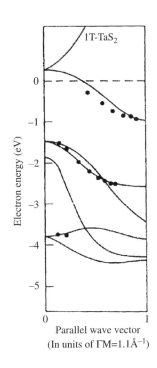

Fig. 8.20 Left: ARPES data from the layered compound 1T-TaS$_2$; the electron energy has been plotted relative to the Fermi energy E_F. The upper edge of each spectrum gives the position of the Fermi energy. The variation of the peak positions with emission angle maps out the bands of the materials; the values of $(E_i, \mathbf{k}_i)$ deduced (points) are plotted alongside a theoretical bandstructure calculation (curves) on the right-hand side of the figure. (After F.J. Himpsel, *Advances in Physics* **32**, 1 (1983).)

compound 1T-TaS$_2$, which is metallic. Note that the upper edge of each spectrum gives the position of the Fermi energy. The variation of the peak positions with emission angle maps out the bands of the materials; the values of $(E_i, \mathbf{k}_i)$ deduced are plotted alongside a theoretical bandstructure calculation in the right-hand side of the figure.

8.7.2 Electroreflectance spectroscopy

In the 1960s and 1970s electroreflectance spectroscopy (ERS) became a widely used technique for the investigation of the electronic band structure of semiconductors. It is a form of 'modulation spectroscopy' and is the most widely used because of the ease with which electric fields may be applied and modulated in a sample. An experiment involves the measurement of the change in the reflectivity of a sample, caused by changing the electric field at the surface.[18] The field is usually applied with a Schottky barrier arrangement, either by evaporating a very thin metallic layer onto the surface of the semiconductor, or by immersing the sample in an electrolyte (the electrolyte is transparent and being a conductor, behaves in an analogous way to the metal).

[18] This is often referred to as the Franz–Keldysh effect. A simple description of the Schottky barrier is given in Section 17.6 of *Electricity and Magnetism*, by B.I. Bleaney and B. Bleaney (OUP, 1989).

When photons interact with a semiconductor, they can cause an electron from an occupied band to be promoted to the conduction band or to a higher unoccupied band. In the case of direct transitions, a 'Joint band diagram' can be drawn in which the energy of the transition is plotted against **k** within the Brillouin zone. Critical points are defined as (i) minima (ii) saddle points and (iii) maxima in this diagram and are designated M_0 for minima, M_1 and M_2 for saddle points, and M_3 for maxima. From the joint band diagram, a **joint density of states** can be calculated. Singularities, known as Van Hove singularities, occur in the joint density of states at the energies of the critical points, and appear as abrupt changes in slope.

As an example, consider Fig. 6.6, which illustrates some of the M_0 critical points for GaAs. These are at the energies of (i) the fundamental band-gap of the material, known as E_0 (arrow from band 4 to band 5 at the Γ point), (ii) the gap between the spin-orbit split-off valence band and the conduction band (from band 2 to band 5, also at the Γ point- arrow not shown for clarity), known as $E_0 + \Delta_0$, and (iii) the gaps closer to the L-point known as E_1 and $E_1 + \Delta_1$ (arrows from bands 4 and 3 respectively to band 5).

For a single pair of bands forming an M_0 critical point, no optical transitions can take place below the critical point energy in the absence of an electric field. However, in the presence of an electric field, the conduction and valence bands are 'tipped over' in real space, allowing states from either band to tunnel into the previously forbidden band-gap. Thus there is now a finite chance of an optical transition below the energy of the band-gap, the probability of which falls off exponentially with energy below the band-gap energy. The resulting change in the reflectivity is typically only about 1 part in 10^4 for reasonable modulation fields, and therefore **phase sensitive detection** techniques are used: an AC field is applied, and a **lock-in amplifier** only responds to changes in light intensity at the photodetector (which collects the reflected light) at the same frequency as the AC field. The reason why ERS is so attractive for studying electronic structure is that the ERS signal is essentially zero, *except* at the critical points. The energies of the critical points may thus be determined precisely, as well as other useful parameters such as the effective mass. Non-modulation techniques, such as simple reflectivity or absorption measurements give fairly featureless, slowly varying spectra in the region of most critical points, making the determination of their energies impossible from such data.

Figure 8.21 shows ERS data for CdTe, a II-VI semiconductor which possesses a bandstructure very similar to that of GaAs.

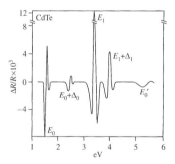

Fig. 8.21 Electroreflectance spectrum of CdTe. Note that the signal is only non-zero at the critical points. (Figure reproduced from M. Cardona, K.L. Shakley and F.H. Pollak, Phys. Rev. **154**, 696 (1967))

8.8 Some case studies

8.8.1 Copper

The Fermi surface of copper was determined in a classic series of experiments involving the techniques in this chapter. Enthusiasts might like to look up the original papers, which are listed below.

- A.B. Pippard *Phil. Trans. Roy. Soc.* **A250** 325 (1957) (skin effect measurements leading to the first good model of the Fermi surface).

- D. Shoenberg *Phil. Trans. Roy. Soc.* **A255** 85 (1962) (de Haas–van Alphen measurements); D.J. Roaf *Phil. Trans. Roy. Soc.* **A255** 135 (1962) (theoretical interpretation of de Haas–van Alphen measurements).

- A.F Kip *et al. Phys. Rev.* **124** 359 (1961); J.F. Koch *et al. Phys. Rev.* **133** A240 (1964) (cyclotron resonance).

- B. Stöhr *et al. Phys. Rev. B* **17** 589 (1979) (angle-resolved photoemission and review of more recent work).

8.8.2 Recent controversy: Sr_2RuO_4

Sr_2RuO_4 is an $n = 1$ member of the family of Ruddlesden–Popper oxides of Ru (see Fig. 7.19). It is of great interest, because it has the same crystal structure as $(La_{1-x}Sr_x)_2CuO_4$, an important member of the high-T_c cuprates. In contrast to the latter materials, however, it has been possible to observe magnetic quantum oscillations in Sr_2RuO_4, and hence make detailed studies of its Fermi surface. The material has a number of other fundamentally interesting properties (e.g. it exhibits an exotic type of superconductivity) and seems likely to continue to be of interest for some time. The material has a relatively simple quasi-two-dimensional Fermi surface, and so the details reported in the papers listed below are followed relatively easily.

In the present context, the studies of Sr_2RuO_4 serve to illustrate the strengths and weaknesses of the various techniques studied in this chapter. ARPES measurements (T. Yokoya *et al., Phys. Rev. Lett.* **76**, 3009 (1996); *Phys. Rev. Lett.* **78**, 2272 (1997); *Phys. Rev. B* **54**, 13311 (1996)) were successful in probing the bands at energies well below E_F and at higher temperatures. However, the Fermi-surface topology implied by the ARPES experiments is clearly at variance with very detailed and unambiguous measurements of de Haas–van Alphen oscillations (A.P. McKenzie *et al., J. Phys. Soc. Jpn.* **67**, 385 (1998), C. Bergemann *et al., Phys. Rev. Lett.* **84**, 2662 (2000).) This discrepancy is almost certainly due to a surface reconstruction (see A. Damascelli *et al., Phys. Rev. Lett.* **85**, 5194 (2000)).

For the enthusiast, general reviews of the properties of Sr_2RuO_4 have been given by Y. Maeno *et al.* in *Nature* **372**, 532 (1994) and *Physica C* **282–287**, 206 (1997).

8.8.3 Studies of the Fermi surface of an organic molecular metal

As the Fermi surfaces of these systems are often very simple (see Section 7.4.3), such experiments are rather easy to understand, and will serve as very visual reminders of the material in Chapter 8.[19]

The inset in Fig. 8.22 shows a section through the first Brillouin zone and Fermi surface of κ-(BEDT-TTF)$_2$Cu(NCS)$_2$ (see Section 7.4.3 for a discussion as to how this is derived).

The main part of Fig. 8.22 shows the resistance of a sample κ-(BEDT-TTF)$_2$Cu(NCS)$_2$ as a function of magnetic field for four different temperatures. Several points are apparent;

---

[19] The experiments described in this section are from N. Harrison *et al., J. Phys.: Condensed Matter* **8**, 5415 (1996) and J. Caulfield *et al., J. Phys.: Condensed Matter* **6**, 2911 (1994).

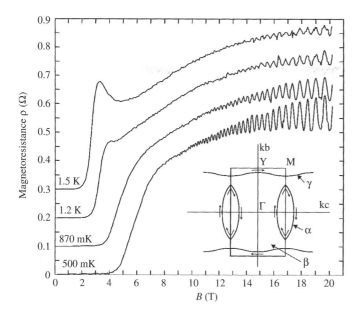

Fig. 8.22 Resistance of κ-(BEDT-TTF)$_2$Cu(NCS)$_2$ as a function of magnetic field for four different temperatures; the data have been offset for clarity (the resistance is zero at $B = 0$ in all cases). The magnetic field has been applied perpendicular to the highly-conducting planes. The inset shows the Brillouin zone, Fermi surface and the notation α, β and γ used to describe the various Fermi-surface orbits.

- the material is a superconductor (its transition temperature is $T_c \approx 10.4$ K); at low magnetic fields it has vanishing resistance, but sufficiently high magnetic fields drive the substance into its 'normal' resistive state;[20]
- Shubnikov–de Haas oscillations can be observed in the resistance;
- the amplitude of the Shubnikov–de Haas oscillations declines with increasing temperature.

[20] See Section H for a brief summary of superconductivity.

The Shubnikov–de Haas oscillations originate from the closed orbits around the hole pocket (labelled α in Fig. 8.22); as described earlier in this chapter, their frequency gives the cross-sectional area of this Fermi-surface section (see eqn 8.32) and the rate at which the oscillations decrease with increasing temperature allows the cyclotron effective mass m^*_{CR} to be estimated (see eqn 8.35). In this case the mass obtained is $m^*_\alpha \approx 3.5 m_e$.

Figure 8.23 shows that the bandstructure is indeed quasi-two-dimensional in nature; resistance data are shown for four orientations of the sample in the magnetic field. As the angle θ between the magnetic field and the normal to the highly-conducting planes increases, features in the resistance (e.g. the Shubnikov–de Haas oscillations) move to higher total magnetic field as $1/\cos\theta$. This is because the Landau quantisation is chiefly determined by the component of the field perpendicular to the two-dimensional (highly-conducting) planes, i.e. $B\cos\theta$. A particular oscillation will therefore occur at a fixed value of $B\cos\theta$, so that it will move to higher total field as θ increases.

Thus far we have used the data to make deductions about the closed hole pocket, labelled α in the inset of Fig. 8.22. Can anything be deduced about the open sheets along the edges of the Brillouin zone? Fig. 8.24 shows a resistance measurement up to very high magnetic fields made using a pulsed-field magnet. At the highest fields, a higher frequency Shubnikov–de Haas oscillation is visible, superimposed on the oscillations due to the hole pocket. These oscillations are the result of magnetic breakdown between the hole

Fig. 8.23 Resistance of κ-(BEDT-TTF)$_2$Cu(NCS)$_2$ as a function of magnetic field at $T = 0.4$ K. Data are shown for four orientations of the sample in the magnetic field; the angle shown is that between the magnetic field and the normal to the highly-conducting planes. The data have been offset for clarity.

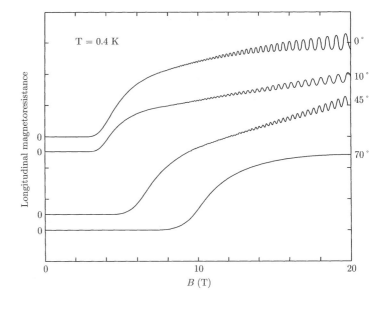

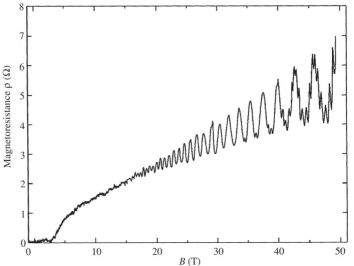

Fig. 8.24 Resistance of κ-(BEDT-TTF)$_2$Cu(NCS)$_2$ as a function of magnetic field at $T = 0.5$ K. The data have been measured using a pulsed magnet; the field was applied perpendicular to the highly-conducting planes.

pocket and the open sheets of the Fermi surface (see Fig. 8.12); as a result, the electrons can carry out orbits around the complete Fermi surface (this is known in the trade as the β orbit). The larger k-space area of the breakdown orbit results in the higher frequency of the Shubnikov–de Haas oscillations.

The frequency of the oscillations yields the k-space area of the complete Fermi surface, which (not surprisingly, given the band filling) is equal to the Brillouin-zone area. The temperature dependence can be used to give a cyclotron mass corresponding to the whole Fermi surface, $m^*_\beta \approx 6.5 m_e$.

We now know an awful lot about the Fermi surface of this material; in the next section, we shall see what happens to the Fermi surface when subtle adjustments are made to the transfer integrals.

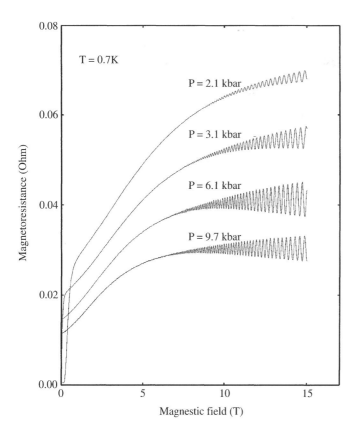

Fig. 8.25 Resistance of κ-(BEDT-TTF)$_2$Cu(NCS)$_2$ at $T = 0.7$ K for several different hydrostatic pressures.

The effect of hydrostatic pressure

Figure 8.25 shows the resistance of κ-(BEDT-TTF)$_2$Cu(NCS)$_2$ at $T = 0.7$ K for several different hydrostatic pressures. A comparison with Fig. 8.22 shows that the increasing pressure suppresses the superconductivity; by $P = 3.1$ kbar, the critical field is a fraction of a Tesla. At 6.1 kbar there is no evidence of superconductivity at all. However, the Shubnikov–de Haas oscillations due to the α hole pocket remain present, so that its k-space area and m_α^* can be extracted as a function of pressure. As the pressure increases, the breakdown oscillations become visible at much lower magnetic fields (cf. Fig. 8.26 and Fig. 8.24), allowing the k-space area of the 'whole' Fermi surface and m_β^* to be deduced.

Figure 8.27 summarises what happens to the Shubnikov–de Haas oscillation frequencies as a function of pressure. The inset shows that the magnetic breakdown frequency increases by about 6 per cent over the pressure range covered. Section 7.4.3 described how the breakdown orbit corresponds to the complete Fermi surface, which has the same cross–sectional area as the Brillouin zone (as there are two holes per unit cell). The breakdown frequency is therefore directly proportional to the k-space area of the Brillouin zone (see eqn 8.32). As the pressure increases, the real-space unit cell will shrink, and therefore the Brillouin zone will get larger (see eqn 2.7), as observed in Fig. 8.27. The measurement of the magnetic breakdown frequency as a function of pressure is therefore in this case a rather obscure way of deriving the compressibility of the material!

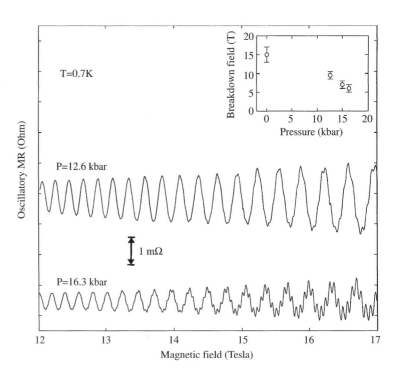

Fig. 8.26 Resistance of κ-(BEDT-TTF)$_2$Cu(NCS)$_2$ at $T = 0.7$ K for two large hydrostatic pressures.

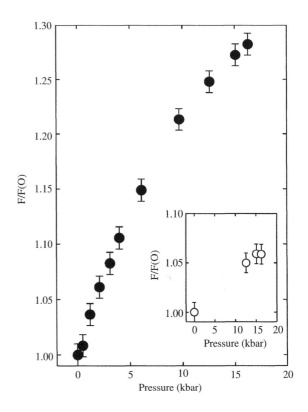

Fig. 8.27 Main figure: the Shubnikov–de Haas oscillation frequency corresponding to the α hole pocket, normalised to its ambient-pressure value, versus pressure. Inset: the pressure dependence of the magnetic breakdown oscillation frequency.

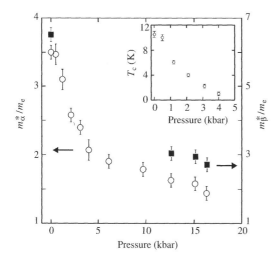

Fig. 8.28 Cyclotron effective masses of the α hole pocket m_α^* and 'whole' Fermi surface m_β^* as a function of pressure (main figure). The inset shows the superconducting critical temperature T_c as a function of pressure.

The main part of Fig. 8.27 shows that over a corresponding range of pressure, the k-space area (directly proportional to the frequency of the Shubnikov–de Haas oscillations; see eqn 8.32) of the α hole pocket changes by a much greater amount. The pressure is changing the transfer integrals which determine the bandstructure, so that the detailed *shape* of the Fermi surface is gradually altering; another symptom of this is the increasing probability of magnetic breakdown (see Fig. 8.26).

The pressure dependence of the masses

Figure 8.28 shows the pressure dependence of m_α^* (corresponding to the α hole pocket), m_β^* (corresponding to the 'whole' Fermi surface) and the superconducting critical temperature.

The application of pressure causes both masses to decrease. This is primarily a consequence of the fact that the BEDT-TTF molecules are pushed closer together so that the transfer integrals increase. As shown in Chapter 4, increasing the transfer integrals results in wider bandwidths and smaller effective masses.

The superconducting critical temperature decreases very rapidly as the masses fall; by 5 kbar, T_c is immeasurably small. In Exercise 1.1 we have seen that the density of states at the Fermi energy is proportional to the effective mass; as the effective masses fall with increasing pressure, the density of states at the Fermi energy will also decrease. Mechanisms for superconductivity involve the pairing of electrons of equal and opposite **k** caused by the exchange of virtual excitations such as phonons or magnetic fluctuations.[21] The strength of the pairing is directly determined by the rate at which this exchange can take place, which in turn depends on the density of states at the Fermi energy. Thus, the increase of the transfer integrals causes the effective masses to decrease, which in turn reduces the density of states at the Fermi energy, thereby suppressing the superconductivity.

Experiments such as the one described in this section are being used to test models of superconductivity. Detailed adjustments to the bandstructure are made using an external variable such as pressure; the changed bandstructure is

[21] See Appendix H and e.g. *Superconductivity* by C.A. Poole, H.A. Farach and R.J. Creswick (Academic Press, San Diego 1995).

then measured using magnetic quantum oscillations. Simultaneous alterations in other properties (e.g. the superconducting T_c) can then be correlated with the bandstructure changes.

8.9 Quasiparticles: interactions between electrons

Measurements of the type described in the previous sections often show that k-space areas of the Fermi-surface sections deduced from the magnetic-quantum oscillation data are in good agreement with the calculated bandstructure. However, if the measured cyclotron effective masses are compared with those predicted from the bandstructure calculations, the agreement is usually much worse.[22]

[22]Taking an example from the previous section, m_α^* in κ-(BEDT-TTF)$_2$Cu(NCS)$_2$ at ambient pressure is $3.5m_e$, whereas the equivalent quantity deduced from bandstructure calculations is $\sim m_e$. In elemental metals, the difference is usually rather smaller; in certain classes of substance, we shall see that the difference can be rather larger.

This is the case in many substances; bandstructure calculations often predict the detailed shape of the Fermi surface rather well but underestimate the observed effective masses. The reason for this can be seen if we consider how we go about a bandstructure calculation; in essence, we constrain ions and molecules to be rigidly fixed in a perfectly periodic arrangement to obtain a periodic potential and hence the bands. However, the ions and/or molecules in a substance will in general be charged, or at the very least possess a dipole moment; as an electron passes through the solid, it will tend to distort the lattice around it owing to the Coulomb interactions between the ions and molecules and its own charge. This leads to the electron being accompanied by a strain field as it moves through the substance. Alternatively one can consider the electron as being surrounded by virtual phonons; the electron and its strain field are often referred to as a 'polaron'. The interactions between the electrons themselves must also be taken into account; as electrons are highly charged, they will repel each other.

By this stage, our initial picture of the electrons as independent particles is beginning to look rather shaky; the electrons interact with each other and carry around a strain field. Landau proposed a way around this problem by supposing that, rather than a system of independent electrons, the solid is populated by a system of independent **quasiparticles**. He then supposed that the quasiparticles obey Pauli's exclusion principle, so that many of the statistical arguments that we have used to describe the electrical properties of solids still work. But what are the quasiparticles? Landau's argument proposes that as electron–electron interactions are turned on, then the states of the strongly interacting electron system evolve continuously from the states of the independent-electron system.

In the case of an independent-electron system, in the derivations of earlier chapters we have (in effect) specified excited states by counting or listing values of **k** (**k**$_1$, **k**$_2$, **k**$_3$, ...**k**$_j$) which denote occupied states with energies above E_F and values of **k** that denote empty states with energies below E_F (**q**$_1$, **q**$_2$, **q**$_3$, ...**q**$_l$). The energy of the excited state is then given by adding the energy corresponding to the states described by **k**$_1$, **k**$_2$, **k**$_3$, ...**k**$_j$ to the ground-state energy and subtracting the energies of the states specified by the **q**$_1$, **q**$_2$, **q**$_3$, ...**q**$_l$. Now in the case of the independent electron system, we would be able to specify the energies corresponding to the **k** and **q** easily (for instance in the nearly-free electron model $E(\mathbf{k}) = \hbar^2\mathbf{k}^2/2m_e$).

In the quasiparticle picture, the excited state may still be defined by saying that l quasiparticles have been excited out of states defined by $\mathbf{q}_1, \mathbf{q}_2, \mathbf{q}_3, \ldots \mathbf{q}_l$ and that j excited quasiparticles have been placed in levels defined by $\mathbf{k}_1, \mathbf{k}_2, \mathbf{k}_3, \ldots \mathbf{k}_j$. The energy of the excited state is still given by adding the energy corresponding to the states described by $\mathbf{k}_1, \mathbf{k}_2, \mathbf{k}_3, \ldots \mathbf{k}_j$ to the ground-state energy and subtracting the energies of the states specified by the $\mathbf{q}_1, \mathbf{q}_2, \mathbf{q}_3, \ldots \mathbf{q}_l$. Thus, the excitations of the interacting system behave in a qualitatively similar way to those of the non-interacting system. However, it is now difficult to determine what the E versus $\mathbf{k}$ relationship of the quasiparticles might be.

The theory describing such quasiparticles is often known as **Fermi liquid theory**; analogously, a non-interacting collection of electrons (as in the Sommerfeld model of Chapter 1) is often referred to as a **Fermi gas**.

For the present purposes, Fermi liquid theory allows one to treat the effect of the various interactions as renormalising the originally calculated effective mass of the independent band electron to give a (usually heavier) quasiparticle effective mass. The quasiparticles then to all intents and purposes behave as band electrons, but with this modified mass. If m_b is the mass calculated in the bandstructure calculations,[23] then the electron–lattice interactions discussed above result in a mass known as the dynamical mass, m_λ, where

$$m_\lambda \approx (1+\lambda)m_b, \qquad (8.42)$$

where λ is an electron–phonon coupling constant (in many substances, λ is between 0.1 and 1).

Returning to the electron–electron interactions, as electrons are highly charged, they will repel each other. Therefore, as an electron is moved across a sample by an external force, there will be a backflow of electrons caused by this repulsion. Thus, it is 'harder' to move electrons about than expected, i.e. the apparent effective mass of the quasiparticles is heavier than m_λ. The effective masses m^* measured in experiments such as the de Haas–van Alphen effect include this contribution from the interactions between the electrons, so that we finally have

$$m^* \approx (1+F)m_\lambda, \qquad (8.43)$$

where F is a constant known as a Fermi liquid parameter.[24]

Thus, both interactions generally act to increase the mass, thus explaining the discrepancy between bandstructure calculations (which generally predict independent band electrons) and experiments (which measure the mass of the quasiparticles). At a simple level, and in systems such as elemental metals, this need not concern us; we can carry on applying models derived for non-interacting electrons, but dealing instead with excitations called quasiparticles which are a little bit heavier, rather than band electrons; all complications are absorbed in the renormalised effective mass.

However, certain classes of materials display very strong interactions, resulting in extremely heavy quasiparticle effective masses. Compounds containing f-electron metals such as UPt_3 and $Ce_2Ru_2Si_2$ are known as **heavy fermion**

---

[23] It is important when comparing measured masses with theoretical predictions to compare like with like. For example, the effective masses described by eqn 5.24 are for linear motion, whereas the masses measured in a magnetic quantum oscillation experiment represent orbital averages; the effective mass deduced from a heat capacity experiment is an average over the whole Fermi surface.

[24] For a detailed treatment of these effects at a reasonably simple level, see Chapters 31–33 of *Many particle theory* by E.K.U. Gross, E. Runge and O. Heinonen (Adam Hilger, Bristol, 1991) or e.g. K.F. Quader *et al.*, *Phys. Rev. B* **36** 156 (1987).

[25] See, for example, S.R. Julian *et al.*, *Physica B* **199–200**, 63 (1994) and references therein.

systems; effective masses as heavy as 100–$1000 m_e$ have been detected in such substances using magnetic quantum oscillations and other techniques.[25]

Further reading

A more detailed treatment of the basic ideas involved in magnetic quantum oscillations is given in *Solid state physics*, by N.W Ashcroft and N.D. Mermin (Holt, Rinehart and Winston, New York, 1976) Chapter 14. Simpler descriptions are in *Electrons in metals and semiconductors*, by R.G. Chambers (Chapman and Hall, London, 1990) Chapter 7, and *Introduction to solid state physics*, by Charles Kittel, seventh edition (Wiley, New York, 1996) Chapters 8 and 9.

For enthusiasts, a splendid and detailed review of magnetic quantum oscillations in metals is found in *Magnetic oscillations in metals*, by David Shoenberg (Cambridge University Press, Cambridge, 1984).

A review of relatively recent cyclotron resonance and interband magneto-optical experiments in semiconductor systems is given in *Optical properties of semiconductors*, Volume 2 of *Handbook on semiconductors*, edited by T.S. Moss and M. Balkanski (Elsevier, Amsterdam 1994), Chapters 6, 7 and 11.

A review of ARPES is given in F.J. Himpsel, *Advances in Physics* **32** 1 (1983); see also M. Grioni *et al.*, *Physica Scripta* **T66**, 172 (1996) and references therein and/or *Fundamentals of semiconductors*, by P. Yu and M. Cardona (Springer, Berlin, 1996), Chapter 8. Further interesting details may be found in articles by E.W. Plummer and W. Eberhardt (*Adv. in Chem. Phys.* **49**, 533 (1982)), and G.J. Lapeyre (*Nucl. Inst. and Meth. in Phys. Research* **A347**, 17 (1994)). The properties of surfaces are reviewed in the book *Electronic properties of surfaces*, edited by M. Prutton (Adam Hilger, Bristol, 1984).

Information on dilution refrigerators and ^{3}He cryostats of the type used for the cryogenic experiments in this chapter is provided in *Experimental low temperature physics*, by A.J. Kent (Macmillan, London 1993), *Matter at low temperatures.*, by P.V.E McLintock, D.J. Meredith and J.K. Wigmore, (Blackie, Glasgow, 1984) and/or *Practical cryogenics; an introduction to laboratory cryogenics* by N.H. Balshaw (published by Oxford Instruments UK Ltd. and available on request).

Reviews of organic molecular metals are given in *Organic superconductors* by T. Ishiguo, K. Yamaji and G. Saito (Springer-Verlag, Berlin 1998), *Fermi surfaces of low-dimensional organic metals and superconductors* by J. Wosnitza (Springer-Verlag, Berlin 1996) and by J. Singleton in *Reports on Progress in Physics* **63** 1111 (2000).

Interacting electron systems and many-body effects are described in the third edition of the excellent book *Many particle physics*, by Gerald Mahan (Kluwer, Dordrecht, 2000) (highly recommended for enthusiasts) or in Many particle theory by E.K.U. Gross, E. Runge and O. Heinonen (Adam Hilger, Bristol, 1991). Two further books which describe classic contributions to this field are *Theory of interacting Fermi systems*, by Philippe Nozieres (Addison Wesley, Reading, 1997), and *Basic notions of condensed matter physics* by P.W. Anderson (Addison Wesley, Reading, 1997).

Heavy fermion systems are reviewed in A.C Hewson, *The Kondo problem to heavy fermions* (Cambridge University Press 1993); E.G. Haanappel, *Physica B* **246–247**, 78 (1998) and references therein. The influence of many-body (i.e. electron–electron and electron–phonon) interactions on the de Haas–van Alphen effect has been reviewed by A. Wasserman and M. Springford, *Advances in Physics* **45**, 471 (1996); this review also contains some references to literature on heavy-fermion systems. A concise summary of recent heavy-fermion data is given in Section 8.2 of *Exotic Kondo effects in metals*, by D.L. Cox and A. Zawadowski (Taylor and Francis, London, 1999). Experimental tests for non-Fermi-liquid behaviour in strongly interacting systems are described in V. Vescoli *et al., Science* **281**, 1181 (1998); C. Bourbonnais and D. Jérome, ibid page 1155 and references therein.

Exercises

(8.1) (a) Figure 8.29 shows $I(9.75)/I(0)$ as a function of photon energy for a sample of a direct-gap semiconductor; the experiment was performed at 77 K. Here $I(B)$ is the intensity of light transmitted by the sample at a magnetic field of B Tesla. Account for the oscillatory behaviour of the plot. Assuming that there are no other strong minima in $I(9.75)/I(0)$ for photon energies below the range shown in the figure, estimate the band gap. Ignore excitonic effects.

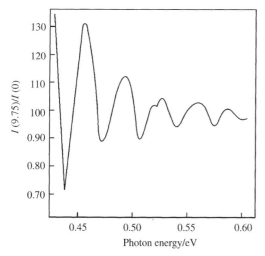

Fig. 8.29 $I(9.75)/I(0)$ as a function of photon energy for a sample of a direct-gap semiconductor; the experiment was performed at 77 K. Here $I(B)$ is the intensity of light transmitted by the InAs sample at a magnetic field of B Tesla.

(b) A sample of the same semiconductor is lightly doped so as to make it n-type. It is found that the donors have a binding energy of 2.1 meV.

Explain why this energy is almost independent of the donor impurity used and use this fact, along with the data in Fig. 8.29, to estimate the effective masses of the electrons and heavy holes (why *heavy* holes and not *light* holes?). (Assume that the bands are isotropic close to the band extrema, so that $m^*_{cCR} \approx m^*_c$ and $m^*_{hCR} \approx m^*_{hh}$. The relative permittivity of the material is $\epsilon_r = 15.2$.)

(8.2) Show that in a magnetic field **B**, the projection of an electron's real-space orbit on a plane perpendicular to **B** is its k-space orbit rotated through $\pi/2$ radians about the field direction and scaled by the factor $\hbar e/B$.

(8.3) Consider the surfaces of constant energy defined by the effective mass tensor of Exercise 5.2. Derive $(\partial A/\partial E)$ for constant energy orbits in planes perpendicular to $\mathbf{e}_3$. Hence show that the cyclotron frequency given by eqns 8.10 and 8.11 for **B** parallel to $\mathbf{e}_3$ is the same as that given by eqn 5.28.

(8.4) Consider the two-dimensional solid of Exercise 3.1. Describe how *electron* and *hole* orbits arise in a magnetic field at certain values of the parameter v, accompanying your answer with sketches.

(8.5) A metal is made up of identical atoms arranged in a crystal with the primitive lattice translation vectors $\mathbf{a} = a\mathbf{e}_1$, $\mathbf{b} = a\mathbf{e}_2$ and $\mathbf{c} = c\mathbf{e}_3$, with $a = 0.3275$ nm and $c = 0.3452$ nm; here $\mathbf{e}_1$, $\mathbf{e}_2$, and $\mathbf{e}_3$ are the Cartesian unit vectors. The low-temperature magnetisation of the crystal is measured with the magnetic field **B** applied parallel to **c**. Two sets of oscillations periodic in $1/B$ are observed in the magnetisation with periodicities $\Delta(1/B) = 3.526 \times 10^{-5} \mathrm{T}^{-1}$ and $\Delta(1/B) = 7.590 \times 10^{-4} \mathrm{T}^{-1}$. However, when the magnetic field is applied parallel to **a** or **b**, no oscillations are seen. Account for these results, and use them to deduce the dimensions and approximate shape of the Fermi surface. Illustrate your answer with a sketch. Is the metal monovalent or divalent?

(8.6) A very pure sample of a metal at a temperature of 1 K is illuminated by microwaves of frequency 72 GHz. At this temperature and frequency the skin depth is less than the electron mean free path. A magnetic flux density **B** is applied parallel to the surface of the sample and at right angles to the electric vector of the radiation. As B is varied the power absorbed by the sample passes through maxima and minima. Maxima are observed when $nB = 0.95$ Tesla, where n is an integer. Account for this behaviour and deduce the electronic dynamical (cyclotron) mass. Estimate the maximum possible error in the alignment of **B** and the sample surface for such an effect to be observed in a typical metal.

(8.7) A nominally undoped piece of semiconductor is illuminated with above-band-gap radiation and a cyclotron absorption measurement is performed at a frequency of 300 GHz and a temperature of 1.5 K. Cyclotron resonances are observed at 1.1 T and 5.8 T; the width (full width at half maximum) of the lower field resonance is 0.2 T and that of the higher field resonance is 1 T. Estimate the masses and mobilities of the carriers induced by the illumination.

How might the measurement be modified to determine which resonance was due to holes and which was due to electrons?

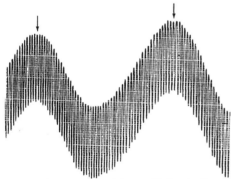

Fig. 8.30 de Haas–van Alphen oscillations in silver.

(8.8) Figure 8.30 shows de Haas–van Alphen oscillations in silver with the magnetic field applied along the [111] direction. Silver has a very similar Fermi surface to that of copper. Explain why two frequencies are seen, and use the data to work out the ratio of the radius of the Fermi-surface neck to the radius of the belly.

(8.9)* A particular monovalent metal has the following properties:

(a) its Fermi surface is approximately spherical;
(b) the Fermi wavevector is $9.1 \times 10^9 \text{m}^{-1}$;
(c) the electrons at the Fermi surface have an effective mass $m^* = 1.3 m_e$, where m_e is the mass of a free electron.

Giving brief experimental details, explain how you would deduce properties (a), (b) and (c) using the **de Haas–van Alphen effect**. Calculate the frequency of the de Haas–van Alphen oscillations that would be observed, the number density of electrons in the metal and its Fermi energy.

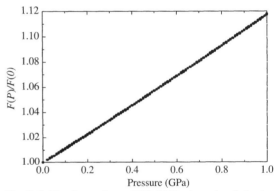

Fig. 8.31 The figure shows $F(P)/F(0)$, the ratio of the de Haas–van Alphen frequency measured with a metal subjected to pressure P to that measured with the metal *in vacuo*.

Figure 8.31 shows $F(P)/F(0)$, the ratio of the de Haas–van Alphen frequency measured with the metal subjected to pressure P to that measured with the metal *in vacuo*. Account for the form of the data, and use the figure to estimate the actual bulk modulus B_{act} of the metal. Calculate the bulk modulus that you would expect for the metal using the Sommerfeld model and comment on your answer.

Describe qualitatively how m^* might be expected to vary with pressure.

Transport of heat and electricity in metals and semiconductors

9.1 A brief digression; life without scattering would be difficult!

In this chapter we shall apply our knowledge of bandstructure, Fermi surfaces and electron statistics to the problem of transport of heat and electricity in metals and semiconductors. To do this, we shall need to reintroduce the idea of a *scattering rate* $1/\tau$ for electrons. Before embarking on detailed derivations, it is instructive to consider what would happen to a band electron if scattering *did not* occur. The result is quite surprising, and readers who are disconcerted by surprises can jump straight to the next section if they wish.

We take as our example the simple bandstructure described in Exercise 5.1; the one-dimensional electron dispersion relationship given is

$$E(k) = E_0 - 2I\cos(ka).$$

We can immediately obtain expressions for the velocity v and effective mass m^* of a band electron using the relationships given in Section 5.2;

$$v \equiv \frac{1}{\hbar}\frac{dE}{dk} = \frac{2Ia}{\hbar}\sin(ka);$$

$$m^* \equiv \frac{\hbar^2}{\frac{d^2E}{dk^2}} = \frac{\hbar^2}{2Ia^2}\frac{1}{\cos(ka)}.$$

The question then states that the band contains just one electron which is at $k = 0$ for $t < 0$; at $t = 0$ an electric field ϵ is turned on in the x direction. For $t \geq 0$, the externally applied force on the electron is $f = -e\epsilon$; the time dependence of k for $t \geq 0$ can therefore be derived using eqn 5.8 to give

$$\hbar\frac{dk}{dt} \equiv f = -e\epsilon.$$

After integration, with the boundary condition $k = 0$ at $t = 0$, we obtain

$$k(t) = -\frac{e\epsilon}{\hbar}t.$$

This can then be substituted in the equations for v and m^* to yield

$$v = -\frac{2Ia}{\hbar}\sin(\frac{a\epsilon et}{\hbar}) \qquad (9.1)$$

9.1	A brief digression; life without scattering would be difficult! 117
9.2	Thermal and electrical conductivity of metals 119
9.3	Electrical conductivity of semiconductors 127
9.4	Disordered systems and hopping conduction 129

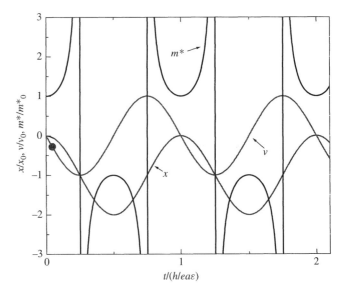

Fig. 9.1 The position x (in units $x_0 = 2I/e\epsilon$), velocity v (in units $v_0 = 2Ia/\hbar$) and effective mass m^* (in units $m_0^* = \hbar^2/2Ia^2$) of the electron discussed in Section 9.1 as a function of time. The point on the velocity curve illustrates the average velocity in the presence of scattering.

and

$$m^* = \frac{\hbar^2}{2Ia^2} \frac{1}{\cos(\frac{a\epsilon et}{\hbar})}, \qquad (9.2)$$

where we have used the identities $\sin(-\theta) = -\sin(\theta)$ and $\cos(-\theta) = \cos(\theta)$ to make the equations look more attractive. Finally, it is a simple matter to integrate the equation for the velocity v with respect to time (setting $x = 0$ at $t = 0$) to give the electron's time-dependent position

$$x = \frac{2I}{e\epsilon}\left[\cos(\frac{a\epsilon et}{\hbar}) - 1\right]. \qquad (9.3)$$

Equations 9.1, 9.2 and 9.3 are plotted in Fig. 9.1; we see the (at first sight) startling result that the electron's position oscillates between $x = 0$ and $x = -2x_0$, where $x_0 = 2I/e\epsilon$, as a function of time! Similarly, the velocity oscillates between $v = \pm v_0 = \pm 2Ia/\hbar$.[1]

[1] This phenomenon is sometimes known as *Bloch oscillations*.

It is useful revision to consider what is going on in more detail. Initially, the electron starts off at the bottom of the band ($k = 0$) with a positive effective mass; it therefore responds to the electric field by moving in the same direction (the negative x direction) as a free, negatively charged electron would. However, the force also changes k uniformly, so that the electron starts to move up the band $E(k)$. As this happens, d^2E/dk^2 decreases; hence the effective mass increases and the acceleration of the electron decreases. Eventually m^* becomes infinite and then flips to $-\infty$ as d^2E/dk^2 goes through zero at $k = -\pi/2a$. An infinite m^* means zero acceleration under an external force; at this point v reaches its maximum negative magnitude. As k changes further, m^* remains negative but becomes smaller in magnitude; the electron's acceleration is now in the *positive* x direction, so that its velocity slows and eventually becomes positive.[2] And so the process goes on; the electron's k continues to change at a uniform rate in time (dk/dt is constant), so that it moves up and down the band $E(k)$. The band is of course periodic in k space, so all of the derived quantities (x, v) also repeat themselves in time.

[2] One could consider that the electron, now near the top of the band, is behaving like a *hole* with positive effective mass and positive charge; its acceleration is therefore in the positive x direction.

As the electron's velocity oscillates about $v = 0$, the electron does not carry any net current in response to the electric field. We therefore arrive at the surprising conclusion that a substance which contains free band electrons acts as an *insulator*![3]

The solution to this peculiar situation is to introduce scattering; in this first discussion, we shall use a very simple form of scattering. Let us suppose that the electron starts to accelerate and then is scattered after a certain time. Let us also suppose that this scattering process, being statistical in nature, completely randomises the electron's k; in other words, the electron 'forgets' everything about its motion before the event. The electron will then start to accelerate again until it scatters once more.

Let us look at the average value of k that might be expected after a scattering event. As the scattering events are random, the electron is as likely to have positive k as it is to have negative k immediately after scattering; statistically (i.e. after averaging many events) the average value of k after scattering will be close to 0. Thus, the average outcome of the scattering is to 'reset' the electron back to its initial state. The effect of the subsequent restricted periods of acceleration will be to give the electron a net, *finite* average velocity in the initial direction of motion. This is represented by the point on the velocity curve in Fig. 9.1.

As the electron has now acquired a finite average velocity, it is carrying a current in response to the field ϵ. Therefore, with the introduction of scattering, the material has become a conductor of electricity. The scattering events prevent the electron moving very far up the band, so that it never accesses the troublesome regions where v and m^* reverse.

In summary, for a band electron to conduct electricity in response to an electric field, some form of scattering must be present. After this rather scary start, I shall introduce some details of the real scattering mechanisms that afflict electrons in solids, and use them to make predictions about the transport of heat and electricity in metals and semiconductors. These predictions describe real substances quite well, and the procedures used in making them are the basis of more complicated and accurate models used in research. We shall start by treating metals.

[3] We have discussed a very simple, one-dimensional band in this section; you should spend a few miniutes convincing yourself that the periodicity of *any* band in k-space (see Section 5.2) will lead to the same result.

9.2 Thermal and electrical conductivity of metals

9.2.1 Metals: the 'Kinetic theory' of electron transport

We are going to look initially at a metal with isotropic bands characterised by an effective mass m^*. This is in effect the Sommerfeld model (quantum mechanical statistics plus free electrons) with the additional refinement of an adjustable m^* which characterises the effect of the periodic lattice potential on the electrons; the Fermi surface will be spherical. (More complex Fermi surfaces will be mentioned in a qualitative manner later.) In Chapter 1 we used the relaxation-time approximation to derive the expression

$$\sigma = \frac{ne^2\tau}{m^*} \quad (9.4)$$

for the electrical conductivity σ of such a metal. Here n is the number density of electrons and τ^{-1} is the scattering rate. In addition, we used kinetic theory and the Sommerfeld model to obtain the expression

$$\kappa = \frac{1}{6} n \pi^2 k_B^2 \frac{T}{E_F} v_F^2 \tau \qquad (9.5)$$

for the electronic thermal conductivity κ. Here v_F is the Fermi velocity and E_F is the Fermi energy. Equations 9.4 and 9.5 were combined to give the Wiedemann–Franz ratio or Lorenz number

$$L \equiv \frac{\kappa}{\sigma T} = \frac{\pi^2}{3} \frac{k_B^2}{e^2} \equiv L_0 \qquad (9.6)$$

and it was found that this constant was in good agreement with many experimental data at room temperature. However, in many cases, experimental values of L differ markedly (by orders of magnitude!) from L_0 at lower temperatures.

The problem has arisen because two scattering times τ (which are not necessarily the same) have been cancelled in eqn 9.6. Acknowledging this fact we rewrite eqns 9.4 and 9.5 in the following form:

$$\sigma = \frac{ne^2 \tau_\sigma}{m^*} \qquad (9.7)$$

and

$$\kappa = \frac{1}{6} n \pi^2 k_B^2 \frac{T}{E_F} v_F^2 \tau_\kappa, \qquad (9.8)$$

i.e. we have distinguished the two scattering times.

9.2.2 What do τ_σ and τ_κ represent?

In all of the following, we are going to be talking about the electron velocity **v**. Previously, we have seen that $\mathbf{v} = (1/\hbar)\nabla_\mathbf{k} E(\mathbf{k})$; this means that electron velocities are always perpendicular to surfaces of constant energy. As always, the electrons at the Fermi surface are going to see most of the action; their velocities may be imagined as vectors normal to the Fermi surface, as the Fermi surface is *the* constant energy surface *par excellence*.[4]

[4] In the case of the simple bandstructure defined at the beginning of this discussion, **v** and **k** *will* be parallel; however, it is worth remembering that this will not usually be the case for bands in general.

Consider first electrical conduction. In the absence of an electric field, there are as many electrons travelling in any one direction as in the opposite direction, so that no net current flows. When an electric field is applied, electrons at the Fermi surface acquire a small amount of extra velocity (the drift velocity derived in the relaxation-time approximation). The effect is shown in Fig. 9.2(a); electrons on the right-hand side of the Fermi surface (with initial velocities parallel to the direction of the drift velocity) move up into slightly higher energy (velocity) states. Other electrons can then fill the states vacated. The net effect is to displace the Fermi surface by a very small amount $\sim m^* v_d/\hbar$, where v_d is the drift velocity (typically $v_d \sim 10^{-3}$ ms^{-1}, whereas $v_F \sim 0.01c$).[5]

[5] See Exercise 1.4.

In order to visualise the equivalent modification of the Fermi surface in the case of thermal conductivity, it is useful to recall the kinetic theory derivation of κ. Thermal conductivity occurs in gases because the gas molecules

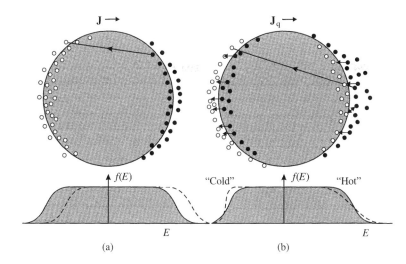

Fig. 9.2 Top: representation of the alteration of the Fermi surface caused by electrical conduction (a) (current density **J**) and thermal conduction (b) (heat flux $\mathbf{J}_q$); black circles represent filled states and white circles empty states. In electrical conduction, the Fermi surface is displaced by a tiny amount corresponding to the drift velocity caused by the electric field; thus, an excess of electrons travel towards the right. In thermal conduction, 'hotter' electrons head right and 'cooler' electrons head left. The effects of these changes are represented in the lower half of the figures by plotting the Fermi–Dirac distribution functions for left-heading and right-heading electrons (dotted lines) compared to the equilibrium situation (grey shading). The long arrows indicate phonon scattering events of the kind encountered at room temperature, which lead to degradation of both the electrical and thermal currents. The short arrows are the phonon scattering processes encountered at low temperatures; note that these can degrade the thermal current both by 'warming up' 'cold' electrons and 'cooling' 'hot' electrons.

travelling in one direction have a higher characteristic thermal energy (or effective temperature) than those travelling in the opposite direction. The equivalent situation for electrons is represented in Fig. 9.2(b); the electrons travelling leftwards are 'cooler' and therefore have a less smeared out Fermi–Dirac distribution function, whereas those going rightwards (in the direction of heat flow) are 'hotter', possessing a more broadened Fermi–Dirac distribution function characteristic of a higher temperature.

We can now identify what τ_κ and τ_σ represent.

- τ_σ represents the characteristic time to randomise/get rid of an electron's excess of forward velocity (or deficiency of backward velocity). A single scattering event of the type shown by an arrow in Fig. 9.2(a) would be sufficient for this purpose; an electron heading rightwards (in the direction of the current flow) has been scattered into an empty state on the opposite side of the Fermi surface with almost opposite momentum and velocity.

- τ_κ represents the characteristic time to randomise/get rid of an electron's excess/deficiency of thermal energy. This can be achieved in a variety of ways shown in Fig. 9.2(b); e.g. an electron can lose $\sim k_\mathrm{B} T$ of energy, and drop into an empty state close by in k-space (vertical process) or it can be shot from the 'hot' side of the Fermi surface to the 'cold' side by a scattering event involving large momentum change (horizontal process).

9.2.3 Matthiessen's rule

In the following discussion, we are going to assume that the electronic scattering rates are additive, i.e.

$$\frac{1}{\tau} = \frac{1}{\tau_1} + \frac{1}{\tau_2} + \frac{1}{\tau_3} + \dots, \tag{9.9}$$

where the τ_j^{-1} are scattering rates due to different processes (e.g. collisions with or emission/absorption of phonons, scattering from impurities etc.). Equation 9.9 implies that the scattering process with the shortest τ_j will dominate, allowing us to predict regions of temperature in which we can ignore all forms of scattering but one; e.g. it is reasonable to assume that scattering of electrons at high temperatures will be almost entirely due to phonons, because there are a lot of phonons around (see Appendix D).

Equation 9.9 is known as **Matthiessen's rule**; it must be admitted that it is only a crude guide as to what to expect in the presence of more than one scattering mechanism. It fails most spectacularly when

- the outcome of one scattering process influences the outcome of another;
- one or more τ_j is a function of **k**.

In the latter case, the conductivities κ and σ will involve total τs due to all processes averaged over all **k**, whereas eqn 9.9 implies the summation of reciprocals of each τ_j individually averaged over all **k**; these two operations are very unlikely to lead to the same result.

Having stated this *caveat*, we shall first think about highish temperatures, where electron–phonon scattering events predominantly determine τ_κ and τ_σ.

9.2.4 Emission and absorption of phonons

When discussing the idea of bands, we have seen that electrons are only scattered when something disturbs the periodicity of the crystal. We are first going to consider the role of phonons, which may be pictured as propagating local distortions of the crystal. Such distortions may scatter an electron; two processes must be considered.

(1) **Elastic processes** Both the phonon and electron change wavevector and energy, constrained by conservation of energy and momentum.

(2) **Inelastic processes** The phonon may be emitted or absorbed by an electron, causing the electron's wavevector and energy to change.

Our derivations of bandstructure also made a hidden assumption, that the positions of the ions were not affected by the presence of mobile electrons. This is of course somewhat unrealistic; electrons and ions are highly charged, and the passage of an electron will result in the distortion of the lattice around it. The electron can be scattered by this; in wave-mechanical terms it has emitted a phonon, causing its energy and momentum to change. This is another inelastic process.

9.2.5 What is the characteristic energy of the phonons involved?

Phonons behave as massless bosons, that is to say they can be created and destroyed in a similar way to photons.[6] Phonons therefore have a 'black-body' type of energy distribution (see Fig. D.7), with a peak at an energy $\sim k_B T$, i.e. the characteristic phonon energy at a temperature T is $\hbar\omega \sim k_B T$. We therefore expect that when an electron scatters from or absorbs a phonon, that phonon will typically have an energy $\sim k_B T$.

But what about the emission of phonons? The form of the Fermi–Dirac distribution function for the electrons means that the only empty states below the Fermi energy have energies $E_F - (\sim k_B T)$. Similarly, the energy of the most energetic electrons will be roughly $E_F + (\sim k_B T)$. Therefore, an electron can only emit a phonon with energy $\sim k_B T$; i.e. even though the electron distorts the lattice at low temperatures, it cannot emit energetic phonons because there are no accessible final states for such a process!

The probability of emitting a phonon of energy $\sim k_B T$ will have a very similar temperature dependence to the probability of absorbing such a phonon. This is easy to see when one considers that the latter depends on the number of phonons around with such an energy (i.e. on the phonon density of states at an energy $\hbar\omega \sim k_B T$) whereas the former depends on the density of available final phonon states (i.e. the phonon density of states at an energy $\hbar\omega \sim k_B T$).

Therefore, in all that follows, I am just going to talk about elastic phonon scattering processes and inelastic phonon scattering processes where the latter means *either* emission *or* absorption of phonons; i.e. I shall assume that the probability of all processes involving phonons will follow a similar temperature dependence.

9.2.6 Electron–phonon scattering at room temperature

The Debye temperatures θ_D of most metals are less than or of the order of room temperature.[7] Now θ_D is roughly the energy of the most energetic phonons in the metal,[8] so that phonons with energy $\hbar\omega \sim k_B T$ will have wavevectors $q \approx$ (half the width of the Brillouin zone) $\sim k_F$, where k_F is the Fermi wavevector. Thus one phonon scattering event (inelastic or elastic) will scatter the electron to the opposite side of the Fermi surface (see Fig. 9.3). Thus $\tau_\sigma^{-1} \approx \tau_\kappa^{-1} \propto$ (number of phonons with $\hbar\omega \sim k_B T$) $\propto T$ (see Appendix D).

9.2.7 Electron–phonon scattering at $T \ll \theta_D$

In this case, phonons with energy $\hbar\omega \sim k_B T$ will have energies $\ll k_B \theta_D$ and therefore $q \ll$ (Brillouin zone size), i.e. $q \ll k_F$. Thus, one *inelastic* phonon scattering event will be able to change the electron's energy by $\sim k_B T$; hence $\tau_\kappa^{-1} \propto$ (number of phonons with $\hbar\omega \sim k_B T$) $\propto T^3$.[9] However, one phonon scattering event (elastic or inelastic) will be unable to knock the electron to the other side of the Fermi surface (see Fig. 9.3) and so $\tau_\sigma \gg \tau_\kappa$. This is the reason for the failure of the Wiedemann–Franz ratio at low temperatures.

In order to take account of the fact that many, many scattering events are required before the excess forward velocity of the electron is thoroughly

[6] See Appendix D or statistical mechanics books such as, for example, *Statistical physics*, by Tony Guenault (Routledge, London, 1988) page 124.

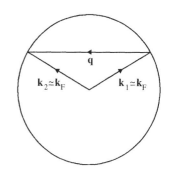

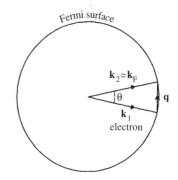

Fig. 9.3 Geometry of electron–phonon scattering events at $T \sim \theta_D$ (above) and $T \ll \theta_D$ (below). In each case $\mathbf{k}_1$ and $\mathbf{k}_2$ are the initial and final electron wavevectors (both of magnitude $\approx k_F$) and $\mathbf{q}$ is the phonon wavevector.

[7] See Appendix D; typical Debye temperatures are given in Table D.1.

[8] See Section D.3.2.

[9] The T^3 power is well known from the Debye heat capacity derivation; see Sections D.3 and D.3.1.

124 *Transport of heat and electricity in metals and semiconductors*

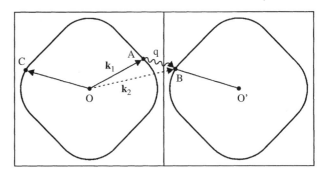

Fig. 9.4 Geometry of electron–phonon umklapp scattering. A phonon with $|\mathbf{q}| < k_F$ is able to scatter an electron of wavevector **k** with a positive velocity from the first Brillouin zone (A) to a state with a negative velocity (B) in the second Brillouin zone. The state at B is equivalent to that at C in the first Brillouin zone.

randomised, the scattering rate contains a weighting factor $(1 - \cos\theta)$, where θ is the scattering angle (see Fig. 9.3). Now as θ is small,

$$1 - \cos\theta \approx 1 - (1 - \frac{\theta^2}{2}) = \frac{\theta^2}{2} \approx \frac{q^2}{2k_F^2} \approx \frac{\omega^2}{2k_F^2 v_\phi^2} \propto T^2, \quad (9.10)$$

where we have used the fact that the dispersion relationship for low energy phonons is close to $\omega = v_\phi q$, with v_ϕ the speed of sound. Therefore we have $\tau_\sigma \propto T^{-5}$ and $\tau_\kappa \propto T^{-3}$.

9.2.8 Departures from the low temperature $\sigma \propto T^{-5}$ dependence

The $\sigma \propto T^{-5}$ temperature dependence is rarely if ever obeyed exactly. There are two main causes involving electron–phonon processes which contribute to this.

- **The periodicity of k-space** often allows phonons with small q to scatter electrons at the Fermi surface into empty states with energy $\sim E_F$ in an adjacent Brillouin zone; these states may have a velocity which is almost opposite to that of the initial state. This is shown schematically in Fig. 9.4; the process is known as **electron umklapp scattering**.
- **Complicated Fermi surfaces** may have lobes, lozenges, ellipsoids etc. (plus their replicas, due to k-space periodicity, from other zones) all over the Brillouin zone. This means that phonons with short q can cause scattering of electrons between Fermi surface sections with very different characteristic velocities.

Both of these effects give a scattering rate which is roughly exponential, $\tau^{-1} \propto e^{-\theta_F/T}$, where θ_F is a characteristic temperature depending on the Fermi-surface geometry.

9.2.9 Very low temperatures and/or very dirty metals

In the case of very low temperatures, the phonon scattering becomes negligible and scattering of electrons by impurities and defects becomes dominant. Impurities have a different ionic core from the host metal, and therefore will

Table 9.1 Summary of the temperature dependences of scattering times and electrical and thermal conductivities.

Temperature (scatterer)	Scattering times	κ σ	Lorenz number
Very low (impurities)	$\tau_\kappa \approx \tau_\sigma$ $\sim$ const	$\kappa \propto T$, $\sigma \sim$ const	L_0
$T \sim \theta_D/10$ (phonons)	$\tau_\kappa \propto T^{-3}$, $\tau_\sigma \propto T^{-5} \to e^{-\theta_F/T}$	$\kappa \propto T^{-2}$, $\sigma \propto T^{-5} \to e^{-\theta_F/T}$	$< L_0$
$T > \sim \theta_D$ (phonons)	$\tau_\kappa \approx \tau_\sigma$ $\propto T^{-1}$	$\kappa =$const, $\sigma \propto T^{-1}$	L_0

often appear to be charged with respect to the background. The scattering of electrons by impurities is therefore rather like Rutherford scattering[10] with electrons being deflected through large angles. One 'event' therefore degrades the transport of heat and electricity equivalently, so that $\tau_\kappa = \tau_\sigma =$ constant and the Wiedemann–Franz ratio again holds.

[10] Rutherford scattering is treated in most 'modern physics' or quantum mechanics texts; see e.g. *Quantum mechanics*, by Stephen Gasiorowicz (Wiley, New York, 1974) page 400.

9.2.10 Summary

A summary of the temperature dependences of scattering times and electrical and thermal conductivities is given in Table 9.1. Typical electrical resistivity and thermal conductivity data are shown in Figs. 9.5 and 9.6. Note that for reasonably pure metals with reasonably simple bandstructures (e.g. alkali and noble metals), electrical resistivity data are all of a similar form, and when normalised to the value at $T = \theta_R$ lie roughly on the same curve when plotted against T/θ_R (see Fig. 9.5); here θ_R is a characteristic temperature similar (but not identical) to θ_D.

9.2.11 Electron–electron scattering

In metals with simple Fermi surfaces, electron–electron scattering is relatively unimportant. Initial and final states for both electrons must have energies close to E_F and wavevectors close in magnitude to k_F; in addition, energy and momentum must be conserved. The combination of these two requirements makes electron–electron scattering quite unlikely (and actually completely forbidden at $T = 0$).

However, electron–electron scattering becomes more important when

- the Fermi surface is complicated, so that the conservation of energy and momentum becomes easy for a wider variety of possible scattering processes and/or

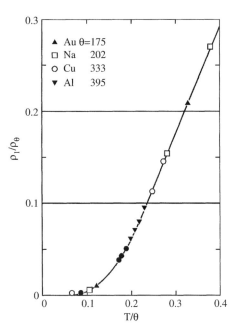

Fig. 9.5 Normalised electrical resistivity data for several metals with reasonably simple Fermi surfaces plotted as a function of the normalised temperature T/θ_R. θ_R is shown in Kelvin for each metal at the top of the figure. Data taken from figures in *Low temperature solid state physics*, by H.M. Rosenberg (OUP, 1963).

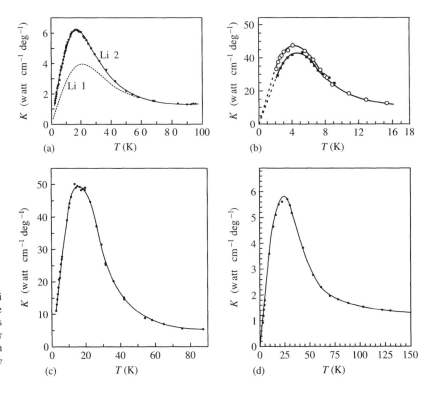

Fig. 9.6 Thermal conductivity data for (a) Li (sample Li 1 is of lower purity than sample Li 2); (b) Na (again showing the effects of dirtier (lower curve) and cleaner (upper curve) samples); (C) Cu; (d) Cr. Data taken from figures in *Low temperature solid state physics*, by H.M. Rosenberg (OUP, 1963).

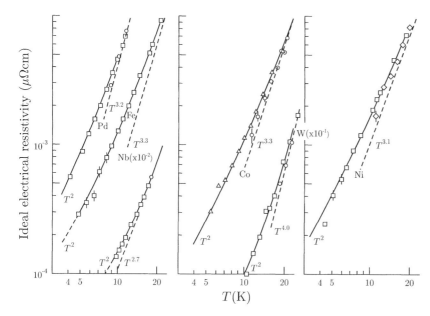

Fig. 9.7 Resistivities of transition metals, showing the approach to T^2 at low temperatures. Data taken from figures in *Low temperature solid state physics*, by H.M. Rosenberg (OUP, 1963).

- the density of states at the Fermi energy is very large (because the effective mass is large), bumping up the number of initial and final states (e.g. transition metals, heavy fermion compounds).

Reasonably simple arguments[11] show that electron–electron scattering leads to $\tau^{-1} \propto T^2$.

[11] See, for example, *Solid state physics*, by N.W Ashcroft and N.D. Mermin (Holt, Rinehart and Winston, New York, 1976) page 347.

Some typical data for transition metals are shown in Fig. 9.7. At low temperatures, the resistivities tend towards the T^2 dependence expected for electron–electron scattering.

9.3 Electrical conductivity of semiconductors

9.3.1 Temperature dependence of the carrier densities

Over wide ranges of temperature, the dominant contribution to the temperature dependence of the electrical conductivity is the rapidly varying number of free carriers. Recalling the results of Section 6.4.4, in order to find n and p when impurities are present we use the law of mass action (eqn 6.14)

$$np = T^3 W e^{-\frac{E_g}{k_B T}}$$

combined with the conservation law (eqn 6.19)

$$n - p = N_D - N_A,$$

where N_D is the density of donors and N_A is the density of acceptors (both are assumed to only provide one carrier each).

The electrical conductivity of a semiconductor consists of a sum of contributions from all carrier types. The simple model of Section 6.4.4, which is a reasonable approximation for many semiconductors, has just two types, electrons and heavy holes. Therefore, the conductivity will contain just two contributions, i.e.

$$\sigma = ne\mu_c + pe\mu_{hh}, \tag{9.11}$$

where μ_c and μ_{hh} are the electron and hole mobilities respectively. The mobilities are defined as the drift velocity of the carrier per unit electric field; the relaxation-time approximation (see the derivation of eqn 1.6) can be used to give

$$\mu_c = \frac{e\tau_c}{m_c^*} \tag{9.12}$$

and

$$\mu_{hh} = \frac{e\tau_{hh}}{m_{hh}^*}, \tag{9.13}$$

where τ_c^{-1} and τ_{hh}^{-1} are the scattering rates for the electrons in the conduction band and the heavy holes in the valence band respectively.

The temperature dependence of the electrical conductivity is therefore determined by convolutions of the temperature dependences of n and τ_c and p and τ_{hh}.

When $n \geq p$ (n-type or intrinsic semiconductors) the holes can be completely ignored in most cases; in the majority of semiconductors, the electrons have a much smaller effective mass than the holes, which also results in a smaller density of states and hence a longer τ. Therefore the electron mobility is often much, much greater than that of the holes.

9.3.2 The temperature dependence of the mobility

There are two important sources of scattering.

(1) **Impurities** ($T \ll \theta_D$). Charged impurity scattering is rather like Rutherford scattering; hence the scattering cross-section varies as E^{-2}. In the nondegenerate case, $E \sim k_B T$, so that the cross-section is proportional to T^{-2}; hence the mean-free path is proportional to T^2. The carrier speed is proportional to $E^{\frac{1}{2}}$, i.e. $T^{\frac{1}{2}}$. Therefore

$$\tau \propto \frac{T^{-\frac{1}{2}}}{T^{-2}} = T^{\frac{3}{2}}. \tag{9.14}$$

Note that this contrasts greatly with the situation in metals ($\tau \sim$ constant for impurity scattering), where all of the action goes on at or close to the Fermi surface; the carriers which scatter from the impurities in metals (see Section 9.2.9) have virtually constant (i.e. temperature-independent) energies. In nondegenerate semiconductors, the free carriers have a quasi-Boltzmann-like energy distribution, so that the average energy of the carriers varies with temperature.

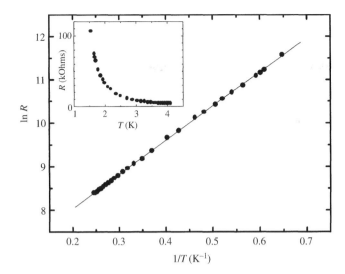

Fig. 9.8 Natural logarithm of the resistance of a carbon resistor (Allen-Bradley 270 Ω) versus $1/T$. The inset shows the resistance versus T. Data are points; the line is a straight-line fit. (Data courtesy of Mervyn Barnes, Oxford Physics Practical Course.)

(2) **Phonons** ($T \sim \theta_\mathrm{D}$). The number of phonons at such temperatures will be proportional to T (see Section 9.2.6), leading to a mean-free-path proportional to T^{-1}. As before, the speed is proportional to $T^{\frac{1}{2}}$. Therefore

$$\tau \propto \frac{T^{-1}}{T^{\frac{1}{2}}} = T^{-\frac{3}{2}}. \tag{9.15}$$

9.4 Disordered systems and hopping conduction

9.4.1 Thermally-activated hopping

Materials which contain large numbers of impurities and defects are often used as low-temperature thermometers. The conductivity of such systems is usually dominated by so-called thermally-activated hopping. At low temperatures, the electrons which carry current through the material, are generally bound in potential wells (resulting from the defects and impurities), known collectively as **traps**. Let us take a simple example, and assume that the traps are an energy Δ deep; in other words, if the electrons are given a thermal energy Δ, they can escape and move through the substance (hence carrying a current) until they next fall into a trap. In such situations, the electrical conductivity σ is roughly proportional to the probability that electrons can escape the traps; from statistical mechanics[12] this probability is proportional to $\exp(-\Delta/k_\mathrm{B}T)$, where T is the temperature. Hence the conductivity is given by

$$\sigma \approx \mathcal{A} e^{-\frac{\Delta}{k_\mathrm{B}T}}, \tag{9.16}$$

where $\mathcal{A}$ is a constant.

Figure 9.8 shows the natural logarithm of the resistance of a carbon resistor (Allen-Bradley 270 Ω) plotted against $1/T$. The data lie on a straight line, showing that the behaviour of the sensor is dominated by a single trap energy over this temperature range.

[12] See, for example, *Statistical physics*, by Tony Guenault (Routledge, London, 1988).

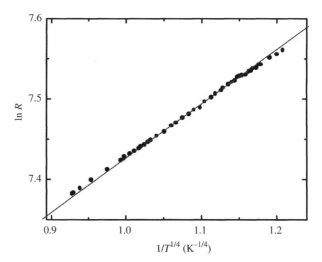

Fig. 9.9 Natural logarithm of the resistance of a Ruthenium oxide resistor (1 kΩ at 290 K) versus $1/T^{\frac{1}{4}}$. Data are points; the line is a straight-line fit. (Data courtesy of Mervyn Barnes, Oxford Physics Practical Course).

9.4.2 Variable range hopping

In materials containing a large number of defects or with a disordered structure the electrons which carry the current are in what is effectively a random potential (spatially the potential energy moves up and down randomly). The electrons gravitate to the potential wells (i.e. dips in the potential) at low temperatures; they can then move through the system by tunnelling (or 'hopping') to adjacent potential wells. This is known as **variable range hopping**, and is of great theoretical interest. The conductivity has a very distinctive temperature dependence[13]

$$\sigma \approx \mathcal{C} \exp(D/T^{\frac{1}{4}}), \tag{9.17}$$

where $\mathcal{C}$ and D are constants. It is the temperature exponent ($\frac{1}{4}$) that identifies this mechanism of conduction.

Figure 9.9 shows the natural logarithm of the resistance of a Ruthenium oxide resistor (1 kΩ at 290 K; these resistors are used as cryogenic thermometers) versus $1/T^{\frac{1}{4}}$. As can be seen from the figure, Ruthenium oxide, which contains large numbers of defects and is generally disordered, obeys eqn 9.17 rather well.

[13] See page 82 of *Matter at low temperatures.*, by P.V.E McLintock, D.J. Meredith and J.K. Wigmore (Blackie, Glasgow, 1984) and references therein for a derivation.

Further reading

There are a large number of similar treatments of this topic e.g. (in increasing order of complexity) *Electricity and magnetism*, by B.I. Bleaney and B. Bleaney, third edition (Oxford University Press, Oxford, 1989) Chapters 11, 17; *Solid state physics*, by G. Burns (Academic Press, Boston, 1995) Sections 9.1–9.14; *Introduction to solid state physics*, by Charles Kittel, seventh edition (Wiley, New York, 1996) Chapters 6 and 7; *Solid state physics*, by N.W Ashcroft and N.D. Mermin (Holt, Rinehart and Winston, New York, 1976) Chapters 1–3, 29.

Those who wish to make calculations of resistivity of a more ambitious nature, will need to use the **Boltzmann transport equation**, which is described in *Solid state physics*, by N.W Ashcroft and N.D. Mermin (Holt, Rinehart and Winston, New York, 1976) pages 316–326. Some tractable applications of the Boltzmann transport equation are worked through in Chapter 7 of *Electron transport in compound semiconductors*, by B.R. Nag (Springer-Verlag, Berlin 1980), and there is a synoptical derivation in Section 7.2 of that book.

More detailed derivations of the various forms of hopping conduction are given in Chapter 1, (Section 15) of *Metal-insulator transitions*, by N.F. Mott (Taylor and Francis, London 1990), which also contains a survey of further literature on these effects. The application of this effect in temperature sensors is discussed in *Matter at low temperatures.*, by P.V.E McLintock, D.J. Meredith and J.K. Wigmore (Blackie, Glasgow, 1984).

In the early days of transport theory, many measurements were made on thin films and wires of comparable size to the mean-free path of the carriers. A lucid review of this early work, with references to key papers, is given in Sections 4.23–4.27 of *Low temperature solid state physics*, by H.M. Rosenberg (OUP 1963).

Exercises

(9.1) An electron in sodium metal is scattered by a phonon. Estimate the typical angle through which the electron is scattered when the temperature T is

(a) $T = 30$ K
(b) $T = 300$ K.

(The relative atomic mass of sodium is 23, its density is 970 kgm^{-3}, and its bulk modulus is 5.2×10^9Nm^{-2}.)
In the light of your answer, explain briefly why the equation

$$\frac{\kappa}{\sigma T} = \eta \left(\frac{k_B}{e}\right)^2, \quad (9.18)$$

where σ is the electrical conductivity, κ is the thermal conductivity, T is the temperature and η is a constant, usually works well at room and at very low temperatures, but fails at intermediate temperatures.

(9.2)* Estimate the ratio of the thermal conductivity due to electrons to that due to phonons at 10^{-2}K for a lead wire 0.5 mm in diameter which has an electrical conductivity of $10^8 \Omega^{-1}$m^{-1}. You may assume the result of the Debye theory (see Appendix D) that the gram molar heat capacity is $1941(T/\theta_D)^3$JK^{-1}gmol^{-1}. (For lead $\theta_D = 108$ K, the bulk modulus is 5.0×10^{10} Pa, the density is 11400 kgm^{-3} and the relative atomic mass is 207.) (This effect is the basis of the 'superconducting switch' used in very low temperature cryogenic experiments.[14])

Why at very low temperatures, is the thermal conductivity of an insulator chiefly limited by the size and shape of the specimen, whereas the thermal conductivity of a metal is usually limited by the purity of the specimen? (Hint: consider the wavelengths of electrons and phonons at low T.)

(9.3)* What is Matthiessen's rule, and under what conditions would you expect appreciable departures from it?

A process sometimes used to purify a metal rod is to pass a large direct curent through it for several days. The current heats the rod and so makes the impurities more mobile; at the same time it exerts a net force on the impurities, through electron–impurity collisions, and so drives them slowly towards one end of the rod. Calculate the net force per unit volume exerted on the impurities in a rod whose residual resistivity (at low temperature) is $10^{-10}\Omega$m, when a current density 10^8Am^{-2} is passed through it. Treat the metal as free-electron-like, with a Fermi energy of 5 eV, and assume that Matthiessen's rule is obeyed.

Would you expect any difference in your result if (a) the carriers were holes or if (b) the carriers had effective masses not equal to m_e?

---

[14] See, for example, *Experimental low temperature physics*, by A.J. Kent (Macmillan, London 1993), or *Matter at low temperatures.*, by P.V.E McLintock, D.J. Meredith and J.K. Wigmore, (Blackie, Glasgow, 1984).

(9.4) Explain briefly why experimental values of $\kappa/(\sigma T)$ often agree with theory better than do the individual values of κ and of σ.

(9.5) Copper can be treated as though it has one conduction electron per copper atom. It is found that when copper is alloyed with 1 atomic per cent of Ge^{4+} the increase in resistivity is $4 \times 10^{-8}\,\Omega\text{m}$. Estimate the corresponding change in resistivity caused by

- 0.1 atomic percent of Ga^{3+} ions;
- 0.1 atomic percent of vacancies.

Lattice distortions may be neglected.

(9.6) Explain the following statements.

- The scattering of electrons from charged impurities at temperatures $T \ll \theta_D$, where θ_D is the Debye temperature, leads to a scattering rate τ^{-1} which is approximately proportional to $T^{-\frac{3}{2}}$ in semiconductors but independent of temperature in metals.
- At temperatures $T >\sim \theta_D$, τ^{-1} is approximately proportional to T in metals and to $T^{\frac{3}{2}}$ in semiconductors.

(9.7) Give an account of the temperature dependence of the electrical conductivity of a lightly-doped (non-degenerate) semiconductor specimen over the temperature range $\sim E_B \leq k_B T \leq \sim E_g/5$, where E_B is the impurity binding energy and E_g is the band gap. Identify the temperatures at which significant changes in scattering mechanism and/or carrier density occur.

(9.8) Using the data from Fig. 9.8, calculate the energy of the traps which dominate the behaviour of the Allen-Bradley (carbon) resistor.

Magnetoresistance in three-dimensional systems

10.1 Introduction

Magnetoresistance is a general term for the changes in the components of the resistivity and conductivity tensors of materials caused by the application of magnetic field. We are going to treat the magnetoresistance of metals in a quite general and simple manner.[1] First, however, the Hall effect in a system with more than one type of carrier will be described, as it helps to illuminate the more general discussion of metals that will follow, and gives a clue as to the origins of magnetoresistance.

10.2 Hall effect with more than one type of carrier

10.2.1 General considerations

We consider the Hall effect with two or more carrier types present (e.g. electrons and holes). The geometry of a Hall effect measurement is shown in Fig. 10.1; the magnetic field $\mathbf{B}$ is applied parallel to the z direction (i.e. $\mathbf{B} = (0, 0, B)$), whilst the current I is driven through the sample in the x direction. The electric field $\mathbf{E}$ is assumed to be $\mathbf{E} = (E_x, E_y, 0)$; we assume that any effect is going to occur in the plane perpendicular to $\mathbf{B}$ because of the nature of the Lorentz force (see eqn 10.1 below). Voltage measuring contacts are provided on the sample so that E_x and E_y can be deduced (see Fig. 10.1).

We assume that the drift velocity $\mathbf{v}$ of each species of carrier can be treated using the relaxation-time approximation, i.e.

$$m^*\{\frac{d\mathbf{v}}{dt} + \frac{\mathbf{v}}{\tau}\} = q\mathbf{E} + q\mathbf{v} \times \mathbf{B}, \quad (10.1)$$

where q is the charge of the carrier, m^* is its effective mass (assumed isotropic and energy-independent) and τ^{-1} is its relaxation (scattering) rate. Note that all changes in $\mathbf{v}$ occur in the plane perpendicular to $\mathbf{B}$; therefore it is sufficient to split eqn 10.1 into x and y components to give

$$m^*\{\frac{dv_x}{dt} + \frac{v_x}{\tau}\} = qE_x + qv_yB \quad (10.2)$$

and

$$m^*\{\frac{dv_y}{dt} + \frac{v_y}{\tau}\} = qE_y - qv_xB. \quad (10.3)$$

10.1	Introduction	133
10.2	Hall effect with more than one type of carrier	133
10.3	Magnetoresistance in metals	135
10.4	The magnetophonon effect	139

[1] In this chapter I shall concentrate on the magnetoresistance which arises due to modification of carrier motion due to the Lorentz force. Magnetoresistance can arise in magnetic conductors for quite different reasons, as has been already mentioned in Section 7.5.

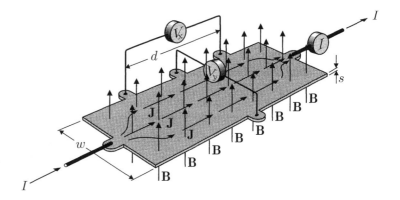

Fig. 10.1 Geometry of a Hall effect measurement on a sample of thickness s and width w. On entering the sample, the current I becomes a current density $\mathbf{J}$ of average magnitude I/ws. The magnetic field (flux density) is uniform within the sample. The positions of voltmeters for measuring $E_x = V_x/d$ and $E_y = V_y/w$ are shown symbolically.

The Hall effect represents a steady state of the system, i.e. $dv_x/dt = dv_y/dt = 0$. Substituting eqn 10.2 into eqn 10.3 with $dv_x/dt = dv_y/dt = 0$ gives

$$\frac{m^* v_y}{\tau} = qE_y - \frac{qB\tau}{m^*}\{qE_x + qv_y B\}, \quad (10.4)$$

which can be rearranged to give

$$v_y\{\frac{m^*}{\tau} + \frac{q^2 B^2 \tau}{m^*}\} = qE_y - \frac{q^2 B\tau}{m^*}E_x. \quad (10.5)$$

Dividing through by m^*/τ and making the identification $eB/m^* \equiv \omega_c$ (i.e. the cyclotron frequency) gives

$$v_y\{1 + \omega_c^2 \tau^2\} = \frac{q\tau}{m^*}\{E_y - \omega_c \tau E_x\}. \quad (10.6)$$

Hall effect experiments are usually quite deliberately carried out at low magnetic fields, such that $\omega_c \tau \ll 1$, implying that terms $\sim \omega_c^2 \tau^2$ can be neglected. Therefore eqn 10.6 becomes

$$v_y = \frac{q\tau}{m^*}\{E_y - \omega_c \tau E_x\} = \frac{q\tau}{m^*}\{E_y - \frac{qB\tau}{m^*}E_x\}. \quad (10.7)$$

We now consider an arbitrary number of carrier types, with each type being labelled by the integer j; the jth carrier type has effective mass m_j^*, charge q_j, scattering rate τ_j and number density n_j. For each carrier type, eqn 10.7 therefore becomes

$$v_{y,j} = \frac{q_j \tau_j}{m_j^*}\{E_y - \frac{q_j B \tau_j}{m_j^*}E_x\}. \quad (10.8)$$

Now the net transverse current must be zero, as there is nowhere for it to go (see Fig. 10.1). Therefore

$$\sum_j n_j v_{y,j} q_j = 0. \quad (10.9)$$

Equations 10.8 and 10.9 can be used to derive the Hall coefficient for an arbitrary number of carrier types. We shall use them to treat the simple case of electrons and heavy holes in a semiconductor; as usual, the light holes, with their relatively feeble density of states compared to that of the heavy holes, will be ignored.

10.2.2 Hall effect in the presence of electrons and holes

In the case of electrons in the conduction band (with effective mass m_c^*, scattering rate τ_c^{-1}, charge q_c, density n) and heavy holes in the valence band (with effective mass m_{hh}^*, scattering rate τ_{hh}^{-1}, charge q_{hh}, density p), eqns 10.8 and 10.9 combine to give

$$\frac{nq_c^2\tau_c}{m_c^*}\{E_y - \frac{q_c B \tau_c}{m_c^*}E_x\} + \frac{pq_{hh}^2\tau_{hh}}{m_{hh}^*}\{E_y - \frac{q_{hh} B \tau_{hh}}{m_{hh}^*}E_x\} = 0. \quad (10.10)$$

Equation 10.10 can be rearranged to give

$$E_y\{n\mu_c + p\mu_{hh}\} = E_x\{p\mu_{hh}^2 - n\mu_c^2\}B, \quad (10.11)$$

where $\mu_{hh} = |q_{hh}\tau_{hh}/m_{hh}^*|$ is the heavy hole mobility and where $\mu_c = |q_c\tau_c/m_c^*|$ is the electron mobility, and I have substituted $q_{hh} \equiv +e$ and $q_c \equiv -e$. Now

$$E_x = \frac{J_x}{\sigma} = \frac{J_x}{|e|(n\mu_c + p\mu_{hh})}, \quad (10.12)$$

where J_x is the current density in the x direction. Combining eqns 10.11 and 10.12 gives

$$R_H \equiv \frac{E_y}{J_x B} = \frac{1}{|e|}\frac{(p\mu_{hh}^2 - n\mu_c^2)}{(n\mu_c + p\mu_{hh})^2}. \quad (10.13)$$

This treatment is explored in more depth in Exercises 10.1 and 10.2.[2]

[2] Some excellent illustrative data are shown in Fig. 4.3 of *Semiconductor physics*, by K. Seeger (Springer, Berlin, 1991).

10.2.3 A clue about the origins of magnetoresistance

Equations 10.9 and 10.10 show that, although no *net* current flows in the y direction, the currents carried in the y-direction by a particular type of carrier may (i.e. probably will) be non-zero. Carriers flowing in the y-direction will experience a Lorentz force caused by **B** in the *negative x-direction* (you can satisfy yourself that this will always be the case). This backflow of carriers will act to change the apparent resistivity E_x/J_x, i.e. cause magnetoresistance.[3]

In the following section we shall explore this idea more formally and in a very general manner. We shall see that the presence of more than one 'carrier type' (and here the term is used very imprecisely) is necessary for magnetoresistance to be observed.

In order to treat all of the different contributions to the conductivity and resistivity in a sanitary fashion, we shall introduce the idea of conductivity and resistivity tensors.

[3] Those who are unconvinced by this handwaving argument should go back to eqn 10.6 and repeat the above derivation for $\omega_c\tau \gg 1$. They will find that $((p/\mu_{hh}) + (n/\mu_c))E_y = (p-n)E_x B$ (i.e the Hall field is zero if $n = p$). Putting $n = p$ yields $J_x = ((p/\mu_{hh}) + (n/\mu_c))eE_x/B^2$, i.e. $\rho \propto B^2$. This is the reason for the very large magnetoresistance in compensated semimetals (equal number of holes and electrons at Fermi surface) such as Bi.

10.3 Magnetoresistance in metals

10.3.1 The absence of magnetoresistance in the Sommerfeld model of metals

We consider first of all a metal with a simple spherical Fermi surface and isotropic, energy-independent effective mass. As in Section 10.2.1, the magnetic field **B** will be parallel to z (see Fig. 10.1), and we shall use the same symbols (effective mass m^*, scattering rate τ^{-1} and electronic charge $-e$). Applying eqn 10.1, we have

$$m^*\left\{\frac{d\mathbf{v}}{dt} + \frac{\mathbf{v}}{\tau}\right\} = -e\mathbf{E} - e\mathbf{v} \times \mathbf{B}. \tag{10.14}$$

Two things may be deduced from this equation.

- The motion of the electrons in the direction parallel to **B** is unaffected. Therefore there will be no longitudinal magnetoresistance; in this context, the longitudinal resistivity is measured in the direction parallel to the field **B** (i.e. both the applied current density and the measured electric field are parallel to **B**).

- There may well be transverse magnetoresistance. Here *transverse* resistivity means that measured in the direction perpendicular to the field **B** (i.e. both the applied current density and the measured electric field are in the plane perpendicular to **B**).

Let us look at the second point in more detail. To simplify matters, we shall initially consider an electric field directed only along the x direction (i.e. $\mathbf{E} = (E_x, 0, 0)$); our tactic will be to deduce the components of the current density $\mathbf{J} = (J_x, J_y, 0)$ that flow in response to **B** and **E**.

As in the previous section, we are dealing with a steady state of the system, i.e. $dv_x/dt = dv_y/dt = 0$. Taking $\mathbf{B} = (0, 0, B)$ and $\mathbf{E} = (E_x, 0, 0)$ as defined above, we rewrite eqn 10.2 and eqn 10.3 in the form

$$v_{d,x} = -\frac{e\tau}{m^*}\{E_x + v_{d,y}B\} \tag{10.15}$$

and

$$v_{d,y} = \frac{e\tau}{m^*}v_{d,x}B, \tag{10.16}$$

where the subscript 'd' emphasises the fact that we are dealing with a drift velocity.

Equations 10.15 and 10.16 show that the magnetic field has made the conductivity anisotropic; it has become a tensor, rather than a scalar. In order to work out the components of the **conductivity tensor**, we look at the current densities $J_x = -nev_{d,x}$ and $J_y = -nev_{d,y}$. Substituting these into eqns 10.15 and 10.16 yields, after some rearrangement (see Exercise 3)

$$J_x = \sigma_{xx}E_x \text{ and } J_y = \sigma_{yx}E_x,$$

where

$$\sigma_{xx} = \frac{\sigma_0}{1 + \omega_c^2\tau^2} \tag{10.17}$$

and

$$\sigma_{yx} = \frac{\sigma_0\omega_c\tau}{1 + \omega_c^2\tau^2}. \tag{10.18}$$

Here $\sigma_0 = ne^2\tau/m^*$ is the zero-field conductivity in the Sommerfeld model and $\omega_c = eB/m^*$ is the cyclotron frequency.

The conductivity tensor shows that, in a magnetic field, the total current density **J** no longer flows parallel to the applied E-field, E_x; instead, it

Fig. 10.2 Geometrical interpretation of the components of current density J_x, J_y and caused by electric field component E_x and magnetic field $(0, 0, B)$; **J** is the total current density.

now contains both x and y components. Figure 10.2 gives a geometrical interpretation of **J** and the components of current density J_x, J_y caused by electric field component E_x and magnetic field $(0, 0, B)$.

Equation 10.17 shows that as $B \to \infty$, $\sigma_{xx} \propto B^{-2}$. We might therefore expect to see some magnetoresistance. However, most experiments (see Fig. 10.1) measure *voltages* dropped in the x- and y-directions between pairs of contacts, rather than measuring the x and y components of the current density. In such experiments the current is forced to go along the x direction, so that $\mathbf{J} \equiv J_x \mathbf{e}_1$; in contrast, the electric field will have components in both x and y directions (see Fig. 10.3). Therefore we want the components

$$\rho_{xx} \equiv \frac{E_x}{J_x} \text{ and } \rho_{yx} \equiv \frac{E_y}{J_x} \tag{10.19}$$

of the *resistivity* tensor, rather than the conductivity.

The general conductivity tensor is

$$\sigma = \begin{pmatrix} \sigma_{xx} & \sigma_{xy} \\ \sigma_{yx} & \sigma_{yy} \end{pmatrix}. \tag{10.20}$$

The derivations which start at eqns 10.15 and 10.16 can be repeated with $\mathbf{E} = (0, E_y, 0)$ to yield $\sigma_{xy} = -\sigma_{yx}$ and $\sigma_{yy} = \sigma_{xx}$ (see Exercise 10.3),[4] so that we have

$$\sigma = \begin{pmatrix} \sigma_{xx} & \sigma_{xy} \\ -\sigma_{xy} & \sigma_{xx} \end{pmatrix} = \frac{\sigma_0}{1+\omega_c^2 \tau^2} \begin{pmatrix} 1 & -\omega_c \tau \\ \omega_c \tau & 1 \end{pmatrix}. \tag{10.21}$$

This tensor can then be inverted using standard methods to give the resistivity tensor

$$\rho = \begin{pmatrix} \rho_{xx} & \rho_{xy} \\ \rho_{yx} & \rho_{yy} \end{pmatrix} = \frac{1}{\sigma_0} \begin{pmatrix} 1 & \omega_c \tau \\ -\omega_c \tau & 1 \end{pmatrix}. \tag{10.22}$$

The components of interest in the experimental arrangement shown in Fig. 10.1 are

$$\rho_{xx} = \rho_0 \text{ and } \rho_{yx} = -\rho_0 \omega_c \tau = -\frac{B}{ne}, \tag{10.23}$$

where $\rho_0 = 1/\sigma_0$ (see Fig. 10.3). Therefore we get no magnetoresistance in the diagonal components of the resistivity tensor and the familiar Hall effect for one carrier in the off-diagonal components.

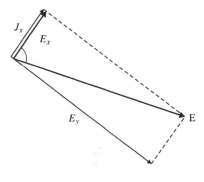

Fig. 10.3 Geometrical interpretation of the components of electric field E_x, E_y and the total field **E** caused by current density component J_x and magnetic field $(0, 0, B)$.

[4] This fact can also be deduced using symmetry considerations.

10.3.2 The presence of magnetoresistance in real metals

Almost all real metals exhibit some form of magnetoresistance, and so we must try to find out what is wrong with the approach above. In the above derivation we assumed that all carriers had the same value of m^* and τ. However, in a real metal we could have

- electrons with different values of m^* (e.g. from anisotropic bands);
- electrons with different values of τ (e.g. some parts of the Fermi surface may have higher scattering probabilities than others);
- a combination of both.

We therefore split the current density **J** into several components

$$\mathbf{J}_j = \sigma_{xx,j} E_x \mathbf{e}_1 + \sigma_{yx,j} E_y \mathbf{e}_2 \tag{10.24}$$

where the index j indicates a contribution from the jth type of carrier. Each type of carrier will have a different density n_j and/or effective mass m_j^* and/or scattering rate τ_j^{-1}. Hence, the components of the conductivity tensor $\sigma_{xx,j}$ and $\sigma_{yx,j}$ will differ for each type of carrier, resulting in $\mathbf{J}_j$s which do not in general point in the same direction as each other.

The total current **J** is just the sum of all of the components

$$\mathbf{J} = \sum_j \mathbf{J}_j. \tag{10.25}$$

In order to see what happens, we take a very simple case of just two carrier types, $j = 1, 2$. Equation 10.24 shows that, barring some very unlikely coincidence, $\mathbf{J}_1$ and $\mathbf{J}_2$ will be in different directions. This situation is illustrated in Fig. 10.4; the application of the magnetic field means that $\mathbf{J}_1$ and $\mathbf{J}_2$ are no longer parallel, so that

$$|\mathbf{J}| \leq |\mathbf{J}_1 + \mathbf{J}_2|, \tag{10.26}$$

i.e. the resistivity increases with increasing magnetic field. We therefore have magnetoresistance.

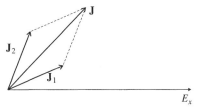

Fig. 10.4 Geometrical interpretation of the components of current density $\mathbf{J}_1$ and $\mathbf{J}_2$ due to two different species of carrier caused by electric field component E_x and magnetic field $(0, 0, B)$.

We note in passing that, as above, $\sigma_{xx} \propto B^{-2}$ as $B \to \infty$ (see eqn 10.17 and the paragraph following it). This will be important in the discussion of the following section.

10.3.3 The use of magnetoresistance in finding the Fermi-surface shape

We consider first a *closed* section of Fermi surface, about which a carrier can perform closed orbits under the influence of a magnetic field (see Fig. 10.5). As $B \to \infty$, $\omega_c \tau \to \infty$, so that an electron will tend to make many circuits of the Fermi surface before scattering. Therefore, the velocity of the electron in the plane perpendicular to **B** will average to zero; this is the reason why σ_{xx} and σ_{yy} both vary as B^{-2} in very high fields. Using the conductivity tensor components, the current densities can be written

$$J_x = \sigma_{xx} E_x + \frac{1}{RB} E_y \tag{10.27}$$

and

$$J_y = -\frac{1}{RB} E_x + \sigma_{yy} E_y \tag{10.28}$$

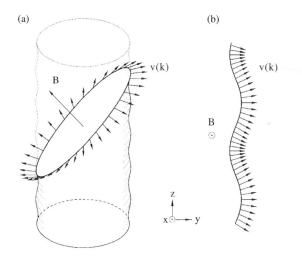

Fig. 10.5 (a) Schematic of electron motion on a closed section of Fermi surface in a magnetic field. The arrows indicate the velocities of an electron following a closed orbit about the Fermi surface in a plane perpendicular to the magnetic field **B**. (b) An open orbit on the Fermi surface. In an in-plane magnetic field, electrons will be driven across the Fermi surface, so that their velocities (shown by arrows) will rock from side to side.

where R is the Hall coefficient, and the off-diagonal tensor components have been written $\sigma_{xy} = 1/RB$ and $\sigma_{yx} = -1/RB$. Eliminating E_y gives

$$E_x = \frac{1}{\sigma_{xx} + (R^2 B^2 \sigma_{yy})^{-1}} J_x + \frac{RB}{1 + R^2 B^2 \sigma_{xx} \sigma_{yy}} J_y. \quad (10.29)$$

As mentioned above, σ_{xx} and σ_{yy} both vary as B^{-2} in very high fields, so that as $B \to \infty$, $(B^2 R^2 \sigma_{yy})^{-1} \gg \sigma_{xx}$. Therefore, $\rho_{xx} = E_x/J_x$ tends to a constant at high fields, i.e. it *saturates*.

We now turn to an *open* section of Fermi surface (see Fig. 10.5) about which an electron cannot perform closed orbits under the influence of a magnetic field. In this case, even as $B \to \infty$, the average value of v_y remains finite; therefore $\sigma_{yy} \to \mathcal{C}$, a constant. Substituting $\sigma_{yy} = \mathcal{C}$ and $\sigma_{xx} = \mathcal{A}B^{-2}$, where $\mathcal{A}$ is another constant, into eqn 10.29 yields

$$\rho_{xx} = \frac{E_x}{J_x} = \frac{1}{\mathcal{A}B^{-2} + (R^2 B^2 \mathcal{C})^{-1}} \propto B^2, \quad (10.30)$$

i.e. $\rho_{xx} \propto B^2$ as $B \to \infty$.

We therefore have two distinct results

- **closed orbits** produce a ρ_{xx} which saturates as $B \to \infty$;
- **open orbits** produce a ρ_{xx} proportional to B^2 as $B \to \infty$.

This has been used to great effect in elucidating the Fermi surface of metals such as copper, where changing the orientation of the magnetic field can produce open or closed orbits about the Fermi surface (see Exercise 10.4).

10.4 The magnetophonon effect

Oscillations can be observed in the resistivity of both bulk and two-dimensional semiconductors at elevated temperatures ~ 100 K; this is known as the **magnetophonon effect** or **magnetophonon resonance**. The effect

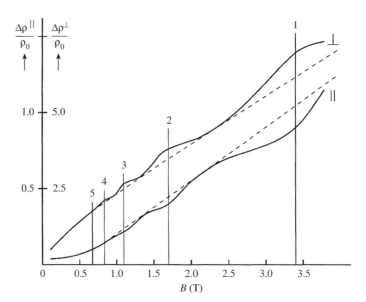

Fig. 10.6 Magnetophonon resonances in the longitudinal and transverse resistivities of InSb at 90 K.

is caused by resonant inter-Landau-level scattering of electrons by long-wavelength longitudinal optic (LO) phonons; such phonons are very effective scatterers of electrons. (Why? think about the type of polarisation field that they produce; see Section D.2.) As such phonons have virtually zero wavevector, the transition is 'vertical', i.e. between almost identical points in k-space in the initial and final Landau levels involved. By conservation of energy, the condition for the magnetophonon effect to occur is therefore

$$j\omega_c = \omega_{LO}, \qquad (10.31)$$

where ω_{LO} is the phonon frequency, i.e. an integer number j of Landau level spacings matches an LO phonon energy. This leads to oscillations in the resistivity periodic in $1/B$; if the phonon frequency is known, the effective mass can be deduced from eqn 10.31.

The conditions for magnetophonon resonance to be observed are

- the temperature should be low enough for the Landau levels to be resolved;
- the temperature should be high enough for a substantial population of LO phonons.

In practice, 70–100 K seems to be a good compromise. Figure 10.6 shows magnetophonon resonances in InSb at 90 K.

Further reading

Some useful general reading on magnetoresistance is contained in *Electrons in metals and semiconductors*, by R.G. Chambers (Chapman and Hall, London, 1990) Chapters 1, 2 and 11, *Semiconductor physics*, by K. Seeger (Springer, Berlin, 1991) Chapter 9 (hard), *Solid state physics*, by N.W Ashcroft and N.D.

Mermin (Holt, Rinehart and Winston, New York, 1976) Chapters 12, 13 and 15 (hard).

Those who wish to calculate magnetoresistance, or who are interested in seeing how the effect helped to determine the Fermi surfaces of a variety of metals are referred to *Magnetoresistance in metals* by A.B. Pippard (Cambridge University Press, 1989).

The angle dependence of the magnetoresistance is of great importance in discovering the Fermi-surface topologies of new materials such as the organic conductors and oxides described in Chapter 7. Recent reviews (including suggestions for tractable calculations based on the Boltzmann transport equation and reasonably up-to-date literature) are given by S.J. Blundell *et al.* (*Phys. Rev. B* **53**, 5609 (1996); *Journal de Physique I*, **6**, 1837 (1996)), A. Ardavan *et al.* (the inclusion of high-frequency effects; *Phys. Rev. B* **60**, 15500 (1999)) and E. Ohmichi *et al.* (application to Sr_2RuO_4; *Phys. Rev. B* **59**, 7263 (1999).

A good review of the magnetophonon effect with relatively recent data is given in *Landau level spectroscopy*, edited by G. Landwehr and E.I. Rashba, Volume 27 of *Modern problems in condensed matter sciences*, edited by V.M. Agranovich and A.A. Maradudin (Elsevier Science Publishers, Amsterdam 1991), Chapters 13 and 20.

Exercises

(10.1) Explain why the electron mobility μ_e is generally much greater than the hole mobility μ_h in many direct-gap semiconductors such as InSb and GaAs.

Figure 10.7 shows R_H plotted as a function of the inverse temperature for a sample of InSb. R_H is observed to change sign between 0.005 and 0.006 K^{-1}. Describe the mechanisms responsible for the form of the curve and estimate (i) the concentration and type of impurities in the sample and (ii) the band-gap of InSb. What temperature dependence would be expected for the mobilities of the carriers over the temperature range shown?

(10.2)* Show that the Hall coefficient R_H of a material containing two types of carrier, with respective concentrations n_1 and n_2, charges q_1 and q_2 and mobilities μ_1 and μ_2 can be expressed as

$$R_H = \frac{1}{e^2} \frac{n_1 q_1 \mu_1^2 + n_2 q_2 \mu_2^2}{(n_1 \mu_1 + n_2 \mu_2)^2}, \quad (10.32)$$

where e is the electronic charge, indicating the approximations inherent in your derivation.

When GaAs is subjected to hydrostatic pressure, the band-gap changes from direct to indirect at a pressure of about 40 kbar, due to the reordering of the conduction band minima at the zone centre and near certain zone boundaries. Describe briefly why the application of pressure affects the shape and width of the bands.

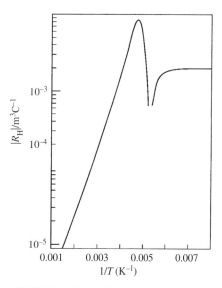

Fig. 10.7 Hall coefficient of an InSb sample versus $1/T$.

Room temperature measurements of (a) the Hall coefficient, normalised to its ambient pressure value, and (b) the Hall mobility, are plotted for n-type GaAs as a function of pressure in Fig. 10.8. What do the data tell you about the conduction band minima involved?

142 *Magnetoresistance in three-dimensional systems*

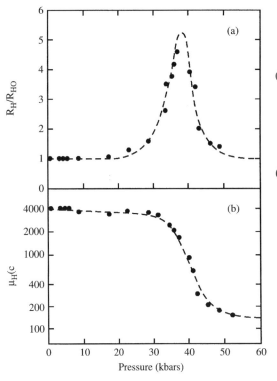

Fig. 10.8 Room temperature measurements of (a) the Hall coefficient, normalised to its ambient pressure value, and (b) the Hall mobility for *n*-type GaAs as a function of pressure.

Use the above formula to show that the peak in the normalised Hall coefficient occurs when $ab \approx 1$, where $a = n_2/n_1$ and $b = \mu_2/\mu_1 \ll 1$. Show that the height of the peak is approximately equal to $(4b)^{-1}$, and compare the value of b deduced this way to that obtained directly from the pressure dependence of the mobility in Fig. 10.8(b). What does this comparison imply about the strength of scattering between the two types of carrier, relative to the strength of scattering between each type of carrier and the lattice?

(10.3) Starting with eqns 10.15 and 10.16 with $\mathbf{E} = (E_x, 0, 0)$, derive the conductivity tensor components σ_{xx} and σ_{yx}. Repeat the derivation using $\mathbf{E} = (0, E_y, 0)$, to show that $\sigma_{yy} = \sigma_{xx}$ and $\sigma_{xy} = -\sigma_{yx}$. Invert the resulting conductivity tensor to find the components of the resistivity tensor.

(10.4) Figure 10.9 shows the magnetoresistance of copper; experimental details are given in the caption. Explain qualitatively the form of the magnetoresistance. (You may find it helpful to refer to Figs. 4.8 and 8.7.)

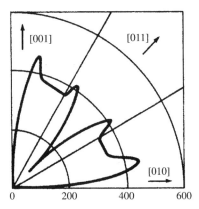

Fig. 10.9 Magnetoresistance of copper at a temperature of 4.2 K and a fixed magnetic field of 1.8 T; the current has been applied in the [100] direction (perpendicular to the plane of the page) and the magnetic field has been rotated from the [001] direction to the [010] direction. The magnetoresistance has been plotted radially as $(\rho(B) - \rho(B = 0))/\rho(B = 0)$. (Data from J.R. Klauder and J.E. Kunzler, *The Fermi surface*, edited by W. Harrison (Wiley, New York, 1960).)

Magnetoresistance in two-dimensional systems and the quantum Hall effect

11.1 Introduction: two-dimensional systems

Chapters 8 and 10 have described the effects of a magnetic field on three-dimensional carrier systems; in particular, Chapter 8 discussed Landau quantisation. We shall now look at two-dimensional semiconductor systems in an analogous manner. The two-dimensionality[1] removes one degree of freedom, so that at high magnetic fields the Landau quantisation results in a density of states consisting entirely of discrete levels, rather than a series of one-dimensional density-of-state functions. This means that the oscillatory phenomena seen in high magnetic fields are often far more extreme in two-dimensional systems than they are in the three-dimensional equivalent.

The reduced dimensionality also results in some other novel phenomena; the quantum Hall effect is a general low-temperature property of two-dimensional or quasi-two-dimensional systems and has been observed in semiconductor structures (e.g. Si MOSFETs, GaAs-(Ga,Al)As heterojunctions and quantum wells containing two-dimensional holes or electrons) and, very recently, in some organic molecular conductors. Most of the research on the quantum Hall effect has been carried out using modulation-doped GaAs-(Ga,Al)As heterojunctions, as these can provide an exceptionally clean two-dimensional electron system with a very low scattering rate (see Chapter 7). We shall therefore confine our discussion to electrons in GaAs-(Ga,Al)As heterojunctions, although similar principles will apply to any of the other systems mentioned above. The effective mass of the electrons in the heterojunction will be labelled m^*, and their areal carrier density N_s.[2]

In a heterojunction, the motion of the electrons in the growth (z) direction is confined by the approximately triangular potential well at the interface; in this direction, the electrons are in bound states of the well known as *subbands*. We let the jth subband have energy E_j. If a magnetic field is applied in the z direction, then the electrons' motion in the xy plane will also become quantised into Landau levels, with energy $(l + \frac{1}{2})\hbar\omega_c$. Therefore, the total energy of the electron becomes

$$E(B, l, j) = (l + \frac{1}{2})\hbar\omega_c + E_j, \qquad (11.1)$$

i.e. the system has become completely quantised into a series of discrete levels.

(We remark at this stage that the magnetic field has reduced the effective dimensionality of the electron system by two; its density of states is now like that of e.g. an atom, with no electronic dispersion whatsoever. This is a general

11.1	Introduction: two-dimensional systems	143
11.2	Two-dimensional Landau-level density of states	144
11.3	Quantisation of the Hall resistivity	147
11.4	Summary	149
11.5	The fractional quantum Hall effect	150
11.6	More than one subband populated	151

[1] The organic conductors and oxides described in Chapter 7 were said to be *quasi-two-dimensional*; although their structures are layered, and most of the electron motion goes on within the layers, there is a finite amount of tunnelling *between* layers, so that the bandstructure is not perfectly two-dimensional. This results in some *warping* of the Fermi surface; for a discussion see J. Singleton, *Reports on progress in physics* **63**, 1111 (2000). By contrast, semiconductor heterojunctions are single layers, and are exactly two dimensional.

[2] It is usual in the literature to use the symbol N_s to denote the number of carriers per unit area, and so I have chosen to do likewise.

Fig. 11.1 The Landau level density of states with zero scattering (left) and in the presence of broadening due to scattering (right). The shaded portions indicate states filled with electrons. Two positions of the chemical potential (Fermi energy) are shown.

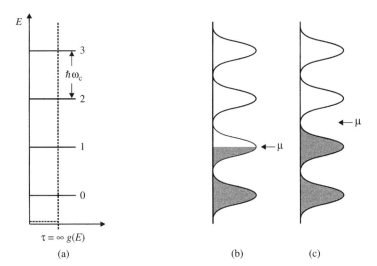

property of a magnetic field; in the case of a three-dimensional system, the electronic density of states becomes a series of density of states functions for one-dimensional motion (see Chapter 9).)

11.2 Two-dimensional Landau-level density of states

Let us assume for the time being that only the lowest subband is populated with electrons. The density of states, which is a series of delta functions separated by $\hbar\omega_c$, is shown in Fig. 11.1(a). In a real sample, these levels will not be perfectly sharp, but instead broadened due to the scattering of electrons by imperfections at the surface of the layer and the presence of impurities. If the typical relaxation time between scattering events is τ (i.e. the lifetime of a particular quantum-mechanical state is τ), the uncertainty principle leads to broadening of the levels by an amount $\delta E \approx \hbar/\tau$. The effects of the magnetic field will therefore only be important if the energy separation of the Landau levels, $\hbar\omega_c$, is larger than this broadening (i.e. $\omega_c\tau \gg 1$). We shall concentrate on this high-field regime of well-separated Landau levels.

Since the number of electronic states at zero field has to equal the number at finite field, the number of electrons which it is possible to put into each Landau level may be calculated from the number of free electron wavevector states with energies between adjacent Landau levels. The wavevector k_l with zero field energy corresponding to the lth Landau level is given by

$$\frac{\hbar^2 k_l^2}{2m^*} = (l + \frac{1}{2})\hbar\omega_c. \tag{11.2}$$

We divide the area of wavevector-space between k_l and k_{l+1} by the well-known degeneracy of two-dimensional phase-space ($N = 2(2\pi)^{-2}V_{r2}V_{k2}$, where V_{r2} is the area in r-space, V_{k2} is the area in k-space and the factor 2 takes into account spin) to get the number of states per unit area per Landau level

$$\frac{N}{V_{r2}} = 2\frac{\pi k_{n+1}^2 - \pi k_n^2}{(2\pi)^2} = \frac{2eB}{h}. \tag{11.3}$$

Each Landau level can hold at most this number of electrons before it is necessary to start to fill the next level.

The electrical properties of the sample will depend very strongly upon how the Landau levels are occupied. Let us assume first of all that the temperature is absolute zero, and that the highest occupied Landau level is half full (Fig. 11.1(b)), i.e. the chemical potential μ is in the middle of a Landau level. Under such conditions, the system looks like a metal; there are lots of empty states which are just above μ, and so the system will have a high electrical conductivity.

Next, consider what happens when the highest filled Landau level is completely filled (Fig. 11.1(c)). In this case, μ will be in the gap between the highest occupied and lowest unoccupied Landau level; there will be no empty states available at energies close to μ. This situation is analogous to that of an insulator; the conductivity of the sample will obviously be a minimum under such circumstances, and for well-separated Landau levels will be effectively zero.

Finite temperatures will cause the Fermi–Dirac distribution function to become 'smeared out' close to μ; however, as long as $\hbar\omega_c >\sim k_B T$, the general picture remains true that the conductivity oscillates dramatically as the field is swept. The conductivity will be very small whenever we have a completely filled highest Landau level, i.e. when

$$N_s = j\frac{2eB}{h}, \tag{11.4}$$

where j is an integer and N_s is the number of electrons per unit area. The maximum conductivity case occurs when

$$N_s = (j + \frac{1}{2})\frac{2eB}{h}. \tag{11.5}$$

The conductivity will therefore oscillate as a function of magnetic field, and will be periodic in $1/B$, with a periodicity given by

$$\Delta(1/B) = \frac{2e}{hN_s}. \tag{11.6}$$

11.2.1 Resistivity and conductivity tensors for a two-dimensional system

The experimental geometry for the quantum Hall effect is usually exactly the same as that for the conventional Hall effect (see Fig. 10.1). Thus, it is the components of the resistivity tensor that we measure, and not the conductivity. We can relate the current densities J_x and J_y (in two dimensions current densities will have the units *current per unit length*) and electric fields E_x and E_y by the expressions

$$J_x = \sigma_{xx}E_x + \sigma_{xy}E_y \tag{11.7}$$

and
$$J_y = -\sigma_{xy}E_x + \sigma_{xx}E_y \tag{11.8}$$

since $\sigma_{xx} = \sigma_{yy}$ and $\sigma_{xy} = -\sigma_{yx}$ for a homgeneous isotropic substance. If, as usual, we allow a current to flow along only one direction (x) (see Fig. 10.1), then $J_y = 0$ and we can say that $E_y/E_x = \sigma_{xy}/\sigma_{xx}$. Therefore we can define the resistivities

$$\rho_{xx} \equiv \frac{E_x}{J_x} = \frac{\sigma_{xx}}{\sigma_{xx}^2 + \sigma_{xy}^2} \tag{11.9}$$

and
$$\rho_{xy} \equiv \frac{E_y}{J_x} = \frac{\sigma_{xy}}{\sigma_{xx}^2 + \sigma_{xy}^2}. \tag{11.10}$$

Note that the units of ρ_{xx} are known as Ω per square; a dimensionless geometrical factor (sample length/sample width) must be introduced to work out the resistance of a particular sample.

We use the relaxation-time approximation to work out the motion of the electrons, i.e.

$$\frac{\partial \mathbf{v}}{\partial t} = -\frac{e}{m^*}\mathbf{E} - \frac{e}{m}\mathbf{v} \times \mathbf{B} - \frac{\mathbf{v}}{\tau}. \tag{11.11}$$

In steady state, the LHS vanishes, and the two components of the equation read

$$v_x = -\frac{e\tau}{m^*}E_x - \omega_c\tau v_y \tag{11.12}$$

and
$$v_y = -\frac{e\tau}{m^*}E_y + \omega_c\tau v_x, \tag{11.13}$$

where we have written $\omega_c = eB/m^*$. If we impose the condition $v_y = 0$ (no current in the y direction) we have

$$\frac{E_y}{E_x} = -\omega_c\tau. \tag{11.14}$$

Writing $J_x = -eN_s v_x$ eqns 11.12 and 11.14 can be combined to give

$$R_H \equiv \frac{E_y}{J_x B} = -\frac{1}{N_s e}. \tag{11.15}$$

R_H is, of course, the *Hall coefficient*. Referring to Fig. 10.1, we see that for a two-dimensional system, $J_x = I/w$ and $E_y = V_y/w$, where w is the width of the sample, i.e.

$$R_H = \frac{V_y}{IB}. \tag{11.16}$$

As in the case of the three-dimensional systems discussed previously, the low-field Hall coefficient can be used to determine N_s.

In high magnetic fields, $\omega_c\tau \gg 1$. Under these conditions, eqn 11.14 indicates that $|\sigma_{xy}| \gg |\sigma_{xx}|$, so that eqns 11.9 and 11.10 give

$$\rho_{xx} \approx \frac{\sigma_{xx}}{\sigma_{xy}^2} \tag{11.17}$$

and
$$\rho_{xy} \approx \frac{1}{\sigma_{xy}} = R_H B. \tag{11.18}$$

We are therefore left with the rather unexpected result that the resistivity is proportional to the conductivity. This is due to the very important influence of the Hall field, and the way in which resistivity and conductivity have been defined; the resistivity is defined for zero Hall current, and the conductivity for zero Hall field.

At the values of magnetic field corresponding to a completely filled highest Landau level, the 'insulating' case, it is the *conductivity* σ_{xx} which goes to zero, and hence the resistivity ρ_{xx} also tends to zero. Physically speaking, what has happened is that, as the probability of scattering goes to zero, an electron (for $E_y = 0$) will go around a cyclotron orbit with a drift in a direction perpendicular to the electric and magnetic fields, and no current will flow along the direction of the applied electric field. For the resistivity (with $J_y = 0$) the Hall field balances out the Lorentz force, and so a current can flow with very small scattering probability, giving a vanishing resistivity. Experimentally, ρ_{xx} can become immeasurably small.

11.3 Quantisation of the Hall resistivity

Some very interesting things happen when we look at the Hall voltage. Experimentally, at the magnetic field values where ρ_{xx} goes to zero, the Hall voltage becomes independent of field, and a **plateau** appears. When ρ_{xx} increases again, then the Hall voltage jumps up to the next plateau. The most interesting feature of this behaviour is the *value* of the Hall resistivity on the plateaux. We may calculate this very easily.

It was mentioned above that ρ_{xx} goes to zero when a whole number of Landau levels are completely filled. At this point the density of electrons is (see eqn 11.4)

$$N_s = j\frac{2eB}{h} \tag{11.19}$$

And therefore the Hall resistivity is given by

$$\rho_{xy} = R_H B = \frac{B}{N_s e} = \frac{1}{j}\frac{h}{2e^2}. \tag{11.20}$$

As we have seen above (see eqn 11.16), for a truly two-dimensional system the width of the sample cancels out (i.e. ρ_{xy} is measured in Ω), so that

$$\rho_{xy} \equiv \frac{V_y}{I} = \frac{1}{j}\frac{h}{2e^2} = \frac{12906.5}{j}\ \Omega. \tag{11.21}$$

The remarkable feature of the experiment is that this value of Hall resistivity, determined only by the fundamental constants h and e, is obtained extremely accurately (see also Section 11.3.2).

However, even though our simple derivation has predicted the value of the quantised resistance, it only gives this value *at one point*, the field at which an integer number of Landau levels are completely filled. In experiments, the plateaux *extend over finite widths of magnetic field*. In order to see why this occurs, we must introduce *disorder*.

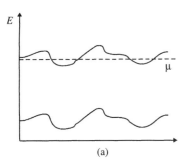

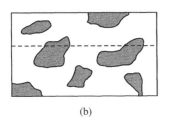

Fig. 11.2 (a) Spatial variation of Landau level energies caused by disorder (impurities etc.). μ indicates the chemical potential. (b) Plan of the resulting 'puddles' of electrons.

11.3.1 Localised and extended states

Figure 11.1 showed broadening of the Landau levels due to scattering of the electrons. One way to look at this is to say that the imperfections in the sample cause a potential energy variation through space, and that the magnitude of this variation is equal to the broadening of the levels. This means that the energy of a Landau level moves up and down as we move through the sample. This is shown in Fig. 11.2(a).

Let us imagine varying the field. Starting at the field at which one Landau level is filled, the field is lowered. This lowers the Landau level degeneracy so that some electrons must go into the next Landau level. These electrons fill up the lowest states, in the minima of the potential fluctuations. The next step is to draw a two-dimensional map of the occupied states. This is shown in Fig. 11.2(b). Provided that the field does not change too much, then the occupied states will be in isolated islands, which are not connected to each other. Because they are isolated, these electrons will not play any part in conduction, and the resistivity and Hall voltage from the sample will remain constant, independent of whether these states are occupied or not. Hence we can see the Hall plateaux and zero σ_{xx} (and ρ_{xx}) over a range of field.

We therefore have two classes of state, **extended states** at the centres of Landau levels, which allow the electrons to move through the heterojunction and **localised states**, in the 'tails' of the Landau levels, which strand the electrons in isolated puddles. When the μ is in the localised states between Landau level centres, $\sigma_{xx} = \rho_{xx} = 0$ and ρ_{xy} is quantised; when μ is in the extended states close to the Landau level centres, σ_{xx} and hence ρ_{xx} are finite.

11.3.2 A further refinement– spin splitting

When experimental data are examined, the Hall plateaux are actually found at resistances

$$\rho_{xy} \equiv \frac{V_y}{I} = \frac{1}{\nu}\frac{h}{e^2} = \frac{25812.8}{\nu}\,\Omega, \tag{11.22}$$

(cf. eqn 11.21) where ν is an integer. The problem is that we have ignored the spin-splitting of the Landau levels in the simple treatment above. The Landau level energies are in fact

$$E(l, B) = \frac{\hbar e B}{m^*}(l + \frac{1}{2}) \pm \frac{1}{2}g^*\mu_B B, \tag{11.23}$$

where μ_B is the Bohr magneton, the $\pm$ takes account of spin-up and spin-down electrons and g^* is the effective g-factor. The resulting Landau-level density of states is shown in Fig. 11.3; we now have twice as many levels as before, each with half the degeneracy, thus explaining the quantised values in eqn 11.22.

The parameter ν is known as the **filling factor**; it is the number of filled (spin-split) half Landau levels, and is given by

$$\nu = \frac{N_s h}{eB}. \tag{11.24}$$

The spin-splitting is much smaller than $\hbar\omega_c$ (e.g. $g^* \sim -0.4$ in GaAs), and so Hall plateaux corresponding to *odd* ν (i.e. those where μ is in a spin gap rather than a cyclotron gap) tend to be resolved only at higher fields.

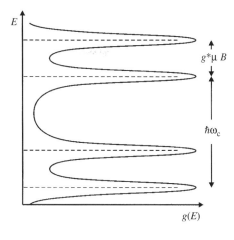

Fig. 11.3 Two-dimensional Landau levels in the presence of spin splitting.

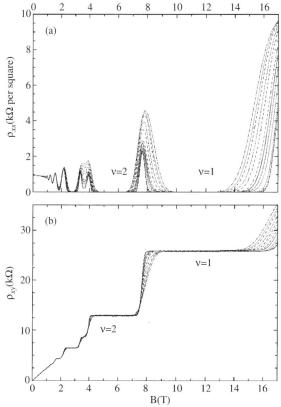

Fig. 11.4 Experimental ρ_{xx} and ρ_{xy} data for a GaAs-(Ga,Al)As heterojunction. Data for temperatures between 30 mK and 1.5 K are shown. (Data courtesy of Phil Gee.)

11.4 Summary

The experimental data in Fig. 11.4 summarise the above discussion very nicely. The following points may be noted.

- At integer ν, the chemical potential μ is in the localised states (i.e. away from the centres of the levels). Therefore $\sigma_{xx} = \rho_{xx} = 0$ and $\rho_{xy} = h/\nu e^2$.

- Between integer ν, μ is in the extended states. Therefore σ_{xx} and ρ_{xx} become finite and ρ_{xy} changes.
- As the temperature rises, the plateaux and regions of zero ρ_{xx} become narrower. This is because the Fermi–Dirac distribution function becomes smeared out on either side of μ, allowing some electrons to access extended states even when μ is amongst the localised states.
- Hall plateaux and ρ_{xx} minima with odd values of ν disappear at low fields. This is because odd ν corresponds to μ being in an energy gap $\sim g^*\mu_B B$, much smaller than $\hbar\omega_c$. By contrast, even ν features correspond to μ being in a gap $\sim \hbar\omega_c$.

The **quantum Hall effect** was first discovered in 1980 by von Klitzing et al.. Von Klitzing was awarded the 1985 Nobel Physics Prize for its discovery. To date the value of the quantized Hall resistance is known to an accuracy of one part in $\sim 10^8$ and has become the international standard for resistance. It is also one of the most accurate ways to determine the atomic fine-structure constant.[3]

[3] The atomic fine structure constant is $\alpha = e^2/(4\pi\epsilon_0 \hbar c)$; α is linked to h/e^2 only by defined constants.

11.5 The fractional quantum Hall effect

Hard on the heels of the discovery of the quantum Hall effect came the observation of the **fractional quantum Hall effect**, for which Stormer, Tsui and Laughlin were awarded the Nobel prize in 1998. In the fractional quantum Hall effect, which is observed only in very high quality samples, additonal plateaux occur at

$$\rho_{xy} \equiv \frac{V_y}{I} = \frac{1}{\nu}\frac{h}{e^2} = \frac{25812.8}{\nu}\,\Omega,$$

with ν taking odd fractional values such as $\frac{1}{3}$, $\frac{2}{3}$, $\frac{1}{5}$, $\frac{2}{5}$ etc.; at the same fields minima are observed in ρ_{xx}. Typical data are shown in Fig. 11.5. The existence of the fractional quantum Hall effect implies that the Landau levels have internal structure.

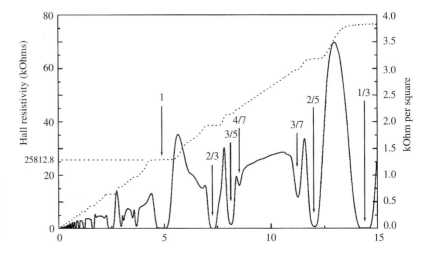

Fig. 11.5 Fractional quantum Hall effect in ρ_{xx} (right-hand scale) and ρ_{xy} (left-hand scale) data for a GaAs-(Ga,Al)As heterojunction ($T = 100$ mK). The filling factors ν are indicated. Data courtesy of Phil Gee, Clarendon Laboratory, Oxford.

Considerable progress in the understanding as to why the particular fractional values of ν occur was made with the introduction of the *composite fermion model*. Note first that the filling factor ν is equal to the inverse of the number of flux quanta per electron (see Exercise 11.4); i.e. $\nu = \frac{1}{2}$ corresponds to 2 flux quanta per electron. In the composite fermion approach, the effects of Coulomb interactions between electrons is replaced by a gauge field, the effect of which is equivalent to attaching $2j$ flux quanta to each electron to form a composite fermion; in other words, the interacting electrons in finite field B are replaced with non-interacting composite fermions in an effective field B^*, given by

$$B^* = B - 2j\Phi_0 N_s, \tag{11.25}$$

where Φ_0 is the flux quantum. Thus for $j = 1$, $B^* = 0$ at $\nu = \frac{1}{2}$.

Notice that the ρ_{xx} data in Fig. 11.5 are rather symmetrical about $\nu = \frac{1}{2}$. The minima in ρ_{xx} close to this field are interpreted as the **integer filling factors of the composite fermions** which occur as the magnitude of the effective field $|B^*|$ increases away from $B^* = 0$ at $\nu = \frac{1}{2}$; just as the electrons in the standard (integer) quantum Hall effect have filling factors $\nu = N_s h/eB$, the composite fermions exhibit Hall plateaux and ρ_{xx} minima at integer **effective filling factor** $\nu^* = N_s h/eB^*$. Substituting for B^* from eqn 11.25, we obtain

$$\nu^* = \frac{N_s h}{e(B - 2jN_s \frac{h}{e})}, \tag{11.26}$$

(where we have substituted the value of the flux quantum) which can be rearranged to give

$$B = \frac{N_s h}{e}\left(2j + \frac{1}{\nu^*}\right). \tag{11.27}$$

Substitution of integer values of ν^* for both positive and negative effective fields in this equation generates all of the observed fractions:

$$\frac{1}{\nu} = 2j + \frac{1}{\nu^*}. \tag{11.28}$$

11.6 More than one subband populated

Figure 11.6 shows ρ_{xx} for a (Ga,In)As-(Al,In)As heterojunction with two populated subbands, plus a schematic representation of the motion of the chemical potential (Fermi energy). At low fields, the levels are poorly resolved, and the effects of spin-splitting are not important; two distinct periodicities are observed, given by

$$\Delta(1/B)_1 = \frac{2e}{hN_{s1}}, \tag{11.29}$$

and

$$\Delta(1/B)_2 = \frac{2e}{hN_{s2}}, \tag{11.30}$$

where N_{s1} and N_{s2} are the populations of the two subbands.[4]

At intermediate fields (~ 10 T), the presence of the second subband complicates the structure of the ρ_{xx} oscillations. However, at high enough fields, the levels will become widely separated in energy and the quantum Hall effect will re-emerge.

[4]This result is a straightforward consequence of the direct proportionality of the cross-sectional area of each subband's two-dimensional Fermi surface to the number of carriers it contains. The discussion of the Shubnikov–de Haas effect in Chapter 8 reminds us that the Fermi-surface area is directly related to the frequency of the oscillations. See also Exercise 11.3.

152 *Magnetoresistance in two-dimensional systems and the quantum Hall effect*

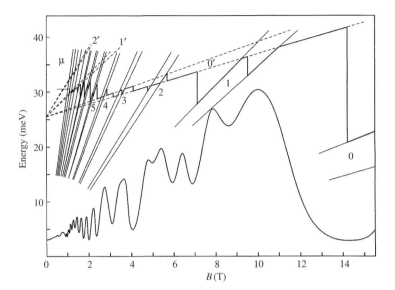

Fig. 11.6 ρ_{xx} for a (Ga,In)As-(Al,In)As heterojunction with two populated subbands, plus a schematic representation of the motion of the chemical potential (Fermi energy). (Data from J.C. Portal *et al.*, *Solid State Commun.* **43**, 907 (1982).)

Further reading

Simple treatments of the quantum Hall effect are found in *Solid state physics*, by G. Burns (Academic Press, Boston, 1995) Chapter 18 and *Low-dimensional semiconductor structures*, by M.J. Kelly (Clarendon Press, Oxford, 1995) (spread through the book).

Harder reviews are provided in *Perspectives in quantum Hall effects; novel quantum liquids in low-dimensional semiconductor structures*, edited by S. Das Sarma and A. Piczuk (Wiley, New York, 1996); *Composite fermions; a unified view of the quantum Hall regime*, edited by O. Heinonen (World Scientific, Singapore, 1998) and *The quantum Hall effect* by R.E. Prange and S.E. Girvin (Springer, Berlin (second edition) 1990); early theoretical approaches have been summarised in Chapter 5 of *Quantum transport in semiconductors*, edited by D.K. Ferry and C. Jacoboni (Plenum, New York, 1991).

Perhaps the most recent review (as this book goes to press) is also the most user-friendly. It is given in 'Fractional quantum Hall effects', by M. Heiblum and A. Stern, *Physics World*, **13**, no. 3, page 37 (2000). An excellent review of the earlier days of the fractional quantum Hall effect is given by J.P. Eisenstein and H.L. Störmer (*Science* **248**, 1510 (1990)). This is worth seeing for its excellent diagrams, and its description of the *Laughlin wavefunction*, the model first used to describe the fractional quantum Hall effect.

Exercises

(11.1) Figure 11.7 shows ρ_{xx} of a GaAs-(Ga,Al)As heterojunction as a function of magnetic field B. The magnetic field is applied perpendicular to the plane of the two-dimensional electron system and data for temperatures from 30 mK to 1.5 K are plotted.

(a) Sketch the electronic Landau level density of states for a heterojunction under such conditions, indicating localised states, extended states and spin splitting. Describe briefly how the changing position of the chemical potential in the Landau levels as a function of magnetic field produces data such as those in Fig. 11.7. Why does increasing the temperature cause the peaks in ρ_{xx} to become broader and larger?

(b) Deduce the areal carrier density and the mean free path length of the electrons (you may assume that the electrons have effective mass $m^* = 0.07 m_e$).

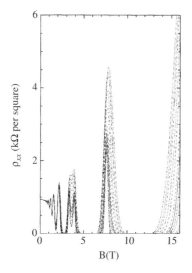

Fig. 11.7 ρ_{xx} for a GaAs heterojunction.

(c) The ρ_{xx} data were recorded by measuring the voltage difference between contacts 1000 μm apart on a 20 μm wide Hall bar along which a current of 10 nA flowed; what voltage would be measured at $B = 0$?

(d) Sketch ρ_{xy} for the same sample at a temperature of 30 mK in as much detail as you can.

(11.2) Make a sketch of the conduction and valence band edges in a GaAs-(Ga,Al)As heterojunction doped so that it contains electrons. Indicate the following items on your sketch; the chemical potential (Fermi energy), the location of the dopants, the spacer layer and the subbands. How does the presence of a spacer layer ensure that the electron mobility is high?

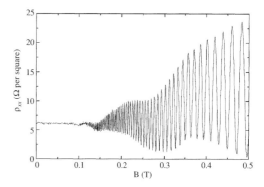

Fig. 11.8 ρ_{xx} for a GaAs-(Ga,Al)As heterojunction at 50 mK.

Figure 11.8 shows ρ_{xx} in a GaAs-(Ga,Al)As heterojunction. Account for the form of the data and estimate the areal carrier density.

(11.3) Show that eqn 11.6 is equivalent to eqn 8.32.

(11.4) Express the magnetic field as number of flux quanta per unit area. By comparing your expression with N_s, the number of electrons pre unit area, show that the filling factor ν is equal to the inverse of the number of flux quanta per electron (i.e. $\nu = \frac{1}{2}$ corresponds to 2 flux quanta per electron).

Use eqn 11.28 to generate the filling factors shown in Fig. 11.5, listing the corresponding values of ν^* and the effective field B^*.

12 Inhomogeneous and hot carrier distributions in semiconductors

12.1 Introduction: inhomogeneous carrier distributions 154
12.2 Drift, diffusion and the Einstein equations 156
12.3 Hot carrier effects and ballistic transport 158

12.1 Introduction: inhomogeneous carrier distributions

In a conventional metal such as Cu, an abnormal distribution of electronic charge will vanish very quickly indeed (in less than about 1 picosec). The reason for this rapid dispersion is of course the strong Coulomb forces between electrons spaced by distances ~ 0.3 nm and the electrons' high mobility. However, in semiconductors, we have the possibility of the simultaneous existence of holes and electrons (see Section 9.3.1). If we inject an excess of *minority* carriers (say holes) into a semiconductor, then the *majority* carriers (say electrons) will move to the region containing the holes to ensure electrical neutrality. After this electrical neutrality is established, there is no longer a strong Coulomb force to disperse the holes, and so the excess of holes will merely diffuse out and/or drift under the action of an external field.

Many electronic devices depend on such non-equilibrium densities of minority carriers. It is not our job to understand the fabrication details and characteristics of these devices in depth; instead we concentrate on the underlying physics of the minority and majority carriers which enables the devices to work.

12.1.1 The excitation of minority carriers

In the following discussion we shall assume that the system is n-type, i.e. the majority carriers are electrons and the minority carriers are holes; the excess densities (over the equilibrium densities) of electrons and holes are labelled Δn and Δp respectively. The discussion can be applied to p-type material by exchanging n for p and 'electrons' for 'holes'.

Minority carriers are usually injected in two ways.

(1) **Above band-gap radiation;** the creation of holes involves the promotion of an equal number of electrons to the conduction band. Therefore $\Delta p = \Delta n$.

(2) **Injection through a contact;** after Δp holes have been injected, $\Delta n \approx \Delta p$ electrons move in extremely quickly to preserve charge neutrality. If they did not, a huge electric field given by $\nabla \cdot \mathbf{E} = e(\Delta p - \Delta n)/\epsilon$ would be present. Therefore $\Delta p \approx \Delta n$ once more.

In other words, an excess of holes is always accompanied by a similar excess of electrons; the notable feature, however, is that

$$\frac{\Delta p}{p} \gg \frac{\Delta n}{n}. \tag{12.1}$$

Therefore *the departure from the equilibrium situation is much more dramatic for minority carriers*; this is the basis of many important devices.

12.1.2 Recombination

The simple recombination of electrons and holes can only take place under stringent conditions of energy and k conservation; e.g. the recombination time for minority holes in very pure n-Ge is $\tau_{hr} \sim 10^{-2}$s, an inordinately long time in solid-state physics. In practice, most recombination occurs via *traps*, i.e. impurities, dangling bonds and surface states, leading to much shorter lifetimes; e.g. an electron gets stuck on a trap and then a hole collides with it.

The recombination rate must be proportional to the number of initial states; it will be of the form apn, where a is a constant. Under equilibrium conditions ($p = p_0, n = n_0$), this must exactly match the rate of thermal generation of electrons and holes. Therefore the rate of generation must be $ap_0 n_0$. After the excitation of excess carriers

$$\frac{dn}{dt} = \frac{dp}{dt} = a(p_0 n_0 - pn). \tag{12.2}$$

According to eqn 12.1, the fractional change in n is very small compared to that in p, so that

$$\frac{dp}{dt} \approx a n_0 (p_0 - p) = -\frac{\Delta p}{\tau_{hr}}. \tag{12.3}$$

This has defined the minority carrier lifetime τ_{hr}.

12.1.3 Diffusion and recombination

Recombination or extraction (by e.g. a contact) can act to diminish Δp locally, leading to diffusion from regions containing more holes. The flux of holes J_{xh} parallel to x is given by

$$J_{xh} = -D_h \frac{\partial p}{\partial x}; \tag{12.4}$$

this defines the hole diffusion coefficient D_h. The net rate of increase of p in thickness δx is therefore given by

$$\frac{\partial}{\partial x}\{-D_h \frac{\partial p}{\partial x}\}\delta x = -D_h \frac{\partial^2 p}{\partial x^2}\delta x. \tag{12.5}$$

In the steady state, this will be equal to the recombination loss rate within δx, given by $((p_0 - p)/\tau_{hr})\delta x$. Therefore

$$\frac{\partial^2 p}{\partial x^2} = \frac{p - p_0}{D_h \tau_{hr}} \tag{12.6}$$

or
$$\frac{\partial^2 \Delta p}{\partial x^2} = \frac{\Delta p}{D_h \tau_{hr}}; \qquad (12.7)$$

whence
$$\Delta p = (\Delta p)_{x=0} e^{-\frac{x}{L_h}}, \qquad (12.8)$$

where $L_h = (D_h \tau_{hr})^{\frac{1}{2}}$ is known as the **hole diffusion length**.

12.2 Drift, diffusion and the Einstein equations

From the previous section, it is clear that the current densities $\mathbf{J}_c$ and $\mathbf{J}_h$ for electrons and holes respectively are made up of two components, a drift current and a diffusion current; i.e.

$$\mathbf{J}_h = \{\mu_h p \mathbf{E} - D_h \nabla p\} e \qquad (12.9)$$

and

$$\mathbf{J}_c = \{-\mu_c n \mathbf{E} - D_c \nabla n\}(-e), \qquad (12.10)$$

where μ_c and μ_h are mobilities and D_c and D_h are diffusion coefficients for the electrons (c) and holes (h).

Let us consider a non-homogeneous situation and let $V = -|\mathbf{E}|x$ represent an electrostatic potential (corresponding to a uniform field). We use eqns 6.11 and 6.13 to represent the effective density of available states in the conduction and valence bands and eqns 6.5 and 6.6 to represent the Fermi–Dirac distribution function, with the extra contribution eV subtracted (added) from (to) the electron (hole) energies, i.e.

$$n(x) \approx N_c e^{-(E_c - eV - \mu)/k_B T} \qquad (12.11)$$

and

$$p(x) \approx N_v e^{-(\mu - E_v + eV)/k_B T}; \qquad (12.12)$$

remember here that the μ in the exponential represents the chemical potential (Fermi level). If the system is in equilibrium, $\mathbf{J}_c$ and $\mathbf{J}_h$ vanish. Substituting eqns 12.11 and 12.12 into eqns 12.9 and 12.10 yields

$$\mu_c = \frac{eD_c}{k_B T}, \quad \mu_h = \frac{eD_h}{k_B T}. \qquad (12.13)$$

These equations are known as **the Einstein equations**.

12.2.1 Characterisation of minority carriers; the Shockley–Haynes experiment

The mobility of minority carriers in semiconductors can be measured directly by a method due originally to Shockley and Haynes. The specimen is cut into a narrow rectangular bar, $\sim 0.5 \times 0.5 \times 10 - 20$ mm^3. An adjustable, steady voltage is applied between the ends to give a field of order 10^3Vm^{-1} along the bar; two electrodes A, B, which function as rectifying contacts, are

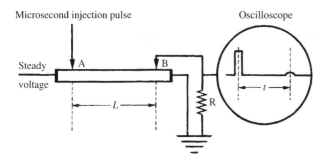

Fig. 12.1 Schematic of the Shockley–Haynes experiment.

applied to the specimen, as shown in Fig. 12.1.[1] A voltage pulse of duration $\sim 1\ \mu$s is applied to electrode A; if the semiconductor is n-type, and A is made positive in the pulse, electrons are withdrawn from the semiconductor by the electrode; some of these may come from the valence band, creating an excess of holes (minority carriers) in the semiconductor. To preserve electrical neutrality, electrons (majority carriers) enter at any contacts which are negative with respect to A. One of these is B, which is connected to a fast data recorder or (through an amplifier) to the y-plates of a storage oscilloscope; this registers a voltage because of the flow of such electrons through the resistance R. This pulse does not quite coincide in time with the pulse at A, but the time delay is that required by an electromagnetic wave set up by the disturbance at A to travel to B, which is $\sim 10^{-4} \mu$s (i.e. negligible). The holes injected at A are swept by the field towards B, and arrive at time $t = L/v$ later, where L is the distance between contacts A and B and v is the drift velocity in the steady field. On arrival at B they appear as a second voltage pulse on the recorder, and the time interval t between the two pulses can be determined. The voltage V between the electrodes A and B due to the steady field is measured independently using a high-impedance voltmeter. Since the drift velocity $v = \mu_{hh} E = \mu_{hh}(V/L)$ and the time $t = L/v$, the mobility μ_{hh} may be found from the equation

$$\mu_{hh} = \frac{L^2}{Vt}. \qquad (12.14)$$

In a typical experiment the value of t is some tens of milliseconds. The second pulse at B due to the arrival of holes is broader and smaller than the first for two reasons:

- the holes diffuse in random directions, broadening the pulse (in some semiconductors this is so rapid that the experiment is impossible);
- holes are lost by recombination with electrons within the specimen.

It is possible to analyse the broadening of the pulse to reveal D_h, the hole diffusion coefficient (see Exercises 12.1–12.3). In order to obtain accurate results, data are recorded for several values of the voltage V, in order to vary the drift velocity and hence transit time t of the pulse between A and B.[2] The pulse arriving at B will be sharp for short transit times, and broad for long transit times, and an analysis of the half-width as a function of t allows an accurate value of D_h to be obtained. Similarly, the total area (proportional to

[1] I shall describe the operation of the experiment for an n-type sample, in which the minority carriers are holes; if a p-type sample (minority carriers electrons) is used, then the signs of the applied voltages are reversed.

[2] An ingenious variant of the apparatus replaces contact A with a focussed light-spot from a flash-tube; the short light pulse generates holes and electrons in equal numbers. The location of the light pulse can be moved along the sample, hence varying L, the distance between the point at which the holes are generated and contact B. This allows the transit time t to be changed in a very controlled manner, and enables errors associated with the exact position of the active part of contact B to be eliminated.

the number of minority carriers) of the pulse can be measured as a function of t, allowing the recombination time τ_{hr} to be deduced.

To measure the mobility of majority carriers (electrons), all that would seem necessary at first sight would be to use n-type material and apply a negative pulse at A. This would inject electrons, creating a local excess, which is, however, dissipated in an extremely short time (see Exercise 12.6), restoring the equilibrium concentration of electrons everywhere in the semiconductor. In this case only the first pulse is observed at B, and no second pulse. When a positive pulse is applied at A, with an n-type semiconductor, holes are created, and although electrical neutrality is restored by an inflow of electrons, we have now a non-equilibrium distribution with an excess of holes and a corresponding excess of electrons. Equilibrium is restored only when the holes flow out at B, or recombine with electrons. The net result is that we can measure directly the mobility only of minority carriers using this technique.[3]

[3] The Hall effect, yielding the carrier density, combined with simultaneous measurements of the resistivity, is generally used to measure the mobility of the majority carriers; see Chapter 10 and Appendix F.

The Shockley–Haynes experiment was very important in the design of the junction transistor. Taking a pnp transistor as an example, the function of the device is determined by the transit of the minority carriers (holes) through the central n-type region. In order for the device to function well, we wish to ensure that any pulse of minority carriers crosses the n-type region quickly (so as not to slow the response of the device down), with minimal broadening due to diffusion (as broadening will also slow the response of the device) and with minimal recombination (as recombination will reduce the efficiency). All of these effects are measured by the Shockley–Haynes experiment, which is nothing more than a very, very stretched pnp transistor; the rectifying contacts A and B correspond to the p-type parts of the junction transistor, and the long bar is the n-type region. It is no coincidence that Shockley was also very much involved in the invention of the junction transistor.

12.3 Hot carrier effects and ballistic transport

12.3.1 Drift velocity saturation and the Gunn effect

Figure 12.2 shows the drift velocity of electrons in GaAs and electrons and holes in Si as a function of electric field.

- In the case of Si, ohmic behaviour persists up to fields of $\sim 10^5 \text{Vm}^{-1}$, with (as has been mentioned *ad nauseam*) the drift velocity of the electrons being higher than that of the holes.

- The drift velocities saturate in Si at fields $\sim 10^6 \text{Vm}^{-1}$. This is because the carriers are losing energy to the lattice by emitting optic phonons, the energy of which is ~ 60 meV in Si. Once a carrier has been accelerated to an energy of ~ 60 meV with respect to the chemical potential, it can emit an optic phonon and lose its excess energy.

- The electrons in GaAs behave in a qualitatively different manner; their velocity first reaches a peak before decreasing to the saturated value.

It is the form of the conduction band in GaAs which is responsible for the distinctive behaviour. The conduction band of GaAs has its lowest point at the zone-centre Γ point, with subsidiary minima, higher in energy, at the X

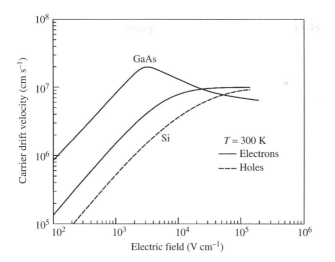

Fig. 12.2 The drift velocity of electrons in GaAs and electrons and holes in Si as a function of electric field.

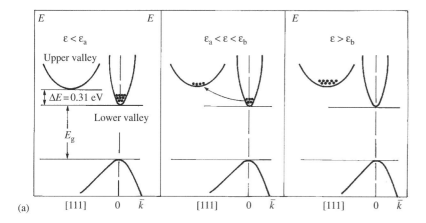

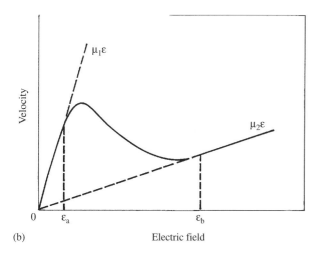

Fig. 12.3 (a) Transfer of carriers between conduction band valleys in GaAs (b) The drift velocity of electrons in GaAs as a function of electric field, showing the regions of high and low mobility and the range of NDC in between. The electric fields ε_a and ε_b in (b) correspond to those mentioned in (a).

and L points at the edge of the Brillouin zone (see Fig. 6.6). The zone-centre minimum is characterised by a relatively light effective mass, whilst the X and L point minima are much less curved, i.e. the effective masses are much heavier. Electrons start off in the Γ point minimum with its low mass and (hence) small density of states; therefore their mobility μ_{c1} is high. As the electric field increases, some of the carriers acquire enough energy to transfer to the X and L minima, where the effective mass and (hence) density of states are higher; their mobility μ_{c2} will be lower. Figure 12.3 shows the resultant velocity field characteristic; at low fields, the velocity increases approximately as $\mu_{c1}|\mathbf{E}|$, whilst at high fields, the velocity increases much more slowly, as $\mu_{c2}|\mathbf{E}|$. The region in between, where the velocity *falls* with increasing $|\mathbf{E}|$ is known as a region of **negative differential conductance** (NDC). NDC is very important in the design of microwave oscillator devices such as the **Gunn diode**.[4]

[4]See *Low-dimensional semiconductor structures*, by M.J. Kelly (Clarendon Press, Oxford, 1995) Chapter 17.

12.3.2 Avalanching

At very high fields $\sim 3 \times 10^7 \mathrm{Vm}^{-1}$, the process shown schematically in Fig. 12.4 takes over. The electrons acquire so much energy that they can relax by creating an electron–hole pair. The resulting electron also gains energy until it too creates another electron–hole pair. This process multiplies the number of carriers and results in **avalanching** of the current. If the process is controlled (by, e.g., operating over very short timescales or having the electric field high only over short distances), it can be used in devices (e.g. the IMPATT oscillator); otherwise it rapidly turns into dielectric breakdown.

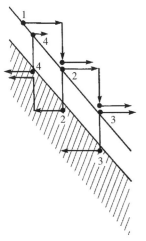

Fig. 12.4 Schematic of avalanching. The diagonal lines represent the conduction-band (upper line) and valence-band (lower line) edges tilted under the influence of a very strong electric field; the shaded region represents the valence band. The electrons acquire so much energy that they can relax by creating an electron–hole pair. The resulting electron also gains energy until it too creates another electron–hole pair.

[5]The label n^+ indicates that a high concentration of donors has been used, so that this region is a degenerate semiconductor; see Sections 6.4.6 and 6.4.7.

12.3.3 A simple resonant tunnelling structure

As has been hinted in Section 12.3.1, NDC is useful in very high frequency oscillators. Using epitaxial growth methods, it is possible to create structures which exhibit strong NDC; the double-barrier resonant tunnelling structure (DBRTS) is the simplest of these.

Figure 12.5 illustrates the principles of the DBRTS. Two (Ga,Al)As layers are embedded in GaAs. The conduction band edges form a one-dimensional double potential barrier with a well between (see Fig. 12.5). The GaAs on either side of the double barriers is n^+ doped[5] (Fig. 12.5A). Applying a bias allows the chemical potential ($\approx E_\mathrm{F}$) in the n^+ region on one side to line up in energy with the quasibound state of the potential well (Fig. 12.5B); the probability of tunnelling into the well will increase dramatically. Once in the well, the carriers can be emitted out of the other side, leading to a large current through the device. As the bias increases, the probability of tunnelling into the well decreases because it is not possible to conserve in-plane wavevector (Fig. 12.5C). The corresponding current-voltage characteristic therefore has pronounced NDC (see Fig. 12.5).

The fall in current in region C is not usually so dramatic in real life as that shown in Fig. 12.5; the presence of phonons enables electrons to get round the problem of in-plane momentum mismatch and to tunnel into the well. Furthermore, charge build-up in the well often occurs, resulting in bistability in and around regions B and C.

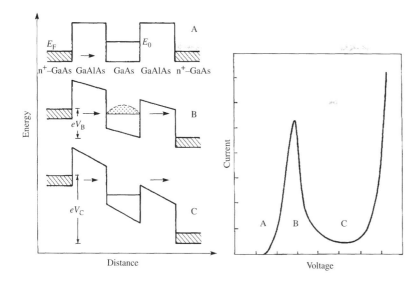

Fig. 12.5 Schematic of DBRTS under various conditions of bias and the corresponding current-voltage characteristic.

12.3.4 Ballistic transport and the quantum point contact

Figure 12.6 illustrates the structure of a quantum point contact; a pair of metallic gates are deposited on top of a semiconductor heterojunction (typically GaAs-(Ga,Al)As) containing two-dimensional electrons. The application of a negative voltage to the gates depletes the electrons underneath, leaving electrons only in a narrow constriction between. A larger negative voltage will 'squeeze' the constriction laterally and further increases will eventually remove the electrons therefrom.

Typical quantum point contact devices are about ~ 0.3 μm long and ~ 0.3 μm wide. This is much less than the typical low-temperature mean-free-path of an electron in a GaAs-(Ga,Al)As heterojunction (see Exercise 11.1); therefore, under the action of an electric field, electrons will usually traverse the constriction **ballistically**, i.e. without scattering. Furthermore, the width of the constriction is such that the electrons' motion perpendicular to the constriction is quantised into one-dimensional subbands, i.e. the electrons are only free to move *through* the constriction; their lateral motion is confined. This has some rather peculiar consequences for the electrical conduction.

Let us imagine applying a bias voltage V across the device in order to make a current flow in the forward direction. As the electrons are effectively one dimensional, the density of states for electrons travelling through the constriction in one particular direction is

$$g(E) = \frac{1}{\pi} \{\frac{m^*}{2\hbar^2}\}^{\frac{1}{2}} E^{-\frac{1}{2}}, \quad (12.15)$$

i.e. half the usual one-dimensional density of states. Now electrons with energy E will have velocity $v = (2E/m^*)^{\frac{1}{2}}$ (we are assuming a parabolic, isotropic band, i.e. constant m^*), so that the current carried in the forward direction by electrons with energy in the range E to $E + \delta E$ will be

$$|\delta J_+| = evg(E)\delta E = e\{\frac{2E}{m^*}\}^{\frac{1}{2}}\frac{1}{\pi}\{\frac{m^*}{2\hbar^2}\}^{\frac{1}{2}}\frac{1}{E^{\frac{1}{2}}}\delta E = \frac{2e}{h}\delta E. \quad (12.16)$$

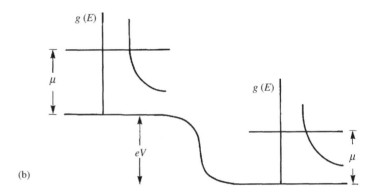

Fig. 12.6 Schematic of quantum point contact: (a) view of the device; (b) schematic of one-dimensional densities of states for electrons moving through the quantum point contact under a bias voltage V.

The total current in the forward direction is therefore

$$|J_+| = \frac{2e}{h}\int_0^\mu dE = \frac{2e\mu}{h}, \tag{12.17}$$

where we have assumed that the highest occupied state has $E = \mu$ (i.e. the temperature is low). However, the total current in the backward direction is

$$|J_-| = \frac{2e}{h}\int_{eV}^\mu dE = \frac{2e(\mu - eV)}{h}, \tag{12.18}$$

where the lower limit includes the effect of the bias voltage (see Fig. 12.6(b)). The net current through the constriction is therefore

$$|J| = |J_+| - |J_-| = \frac{2e^2}{h}V, \tag{12.19}$$

i.e. the conductance is

$$\sigma = \frac{2e^2}{h}. \tag{12.20}$$

If a number j of one-dimensional subbands is occupied, then all will contribute in parallel to the conductance, so that

$$\sigma = \frac{2je^2}{h}, \tag{12.21}$$

i.e. the conductance is completely quantised (note the quantum is the same as that of σ_{xy} in the quantum Hall effect).

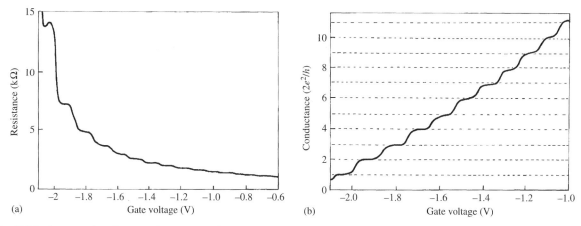

Fig. 12.7 Resistance and conductance of a quantum point contact at 30 mK as a function of gate voltage; note the quantised conductance plateaux. Data taken from B.J. van Wees *et al.*, *Phys. Rev. Lett.* **60**, 848 (1988).

Figure 12.7 shows typical experimental data for a quantum point contact at 30 mK. As the gate voltage becomes more negative, the constriction becomes narrower and the one-dimensional subbands pop through the chemical potential one by one. As each depopulates, the conductivity decreases by $2e^2/h$.

Further reading

General reading on the topic of inhomogeneous carrier distributions will be found in *Solid state physics*, by N.W Ashcroft and N.D. Mermin (Holt, Rinehart and Winston, New York, 1976), Chapter 29 (hard) and *Electricity and magnetism*, by B.I. Bleaney and B. Bleaney, third edition (Oxford University Press, Oxford, 1989), Chapter 17 (easy). Hot electron effects are covered in *Low-dimensional semiconductor structures*, by M.J. Kelly (Clarendon Press, Oxford, 1995), Chapters 1, 4, 6, 7,8, 17 (advanced) and, for example, *Electricity and magnetism*, by B.I. Bleaney and B. Bleaney, third edition (Oxford University Press, Oxford, 1989), Chapter 17 (elementary).

Semiconductor devices are reviewed in *Modern semiconductor device physics* (1997), *Physics of semiconductor devices* (second edition, 1981) and *Semiconductor devices: physics and technology* (1985), all by S.M. Sze (Wiley, New York).

Quantum semiconductor structures by Claude Weisbuch and Borge Winter (Academic Press, San Diego, 1991) provides a very good summary of hetero- and nanostructures at an intermediate level. Some of the theoretical issues involved in mesoscopic devices and nanostructures are covered in *Quantum transport in semiconductors*, edited by D.K. Ferry and C. Jacoboni (Plenum, New York, 1991) and Chapter 5 of *Condensed matter physics* by A. Isihara (OUP, 1991). *Quantum wells, wires and dots: theoretical and computational physics*, by Paul Harrison (Wiley, Chichester, 1999) is an excellent point of departure if you want to carry out calculations and simulations.

Resonant tunnelling diodes have been described in great detail in *The physics and applications of resonant tunnelling diodes*, by H. Mizuta and T. Tanoue (Cambridge University Press, 1995).

Exercises

(12.1) In a Shockley–Haynes experiment on an n-type semiconductor, the equations describing the hole carrier density p and the hole current density J_p are

$$J_p = ep\mu_p E - eD_p \frac{\partial p}{\partial x} \quad (12.22)$$

and

$$\frac{\partial p}{\partial t} = -\frac{1}{e}\frac{\partial J_p}{\partial x} - \frac{p - p_0}{\tau_p}. \quad (12.23)$$

Here, p_0 is the equilibrium hole density. Explain the meaning of the other parameters and the origin of the various terms. Derive a relationship between D_p and μ_p.

(12.2) Assuming that τ_p is large, verify that a solution of eqns 12.22 and 12.23 is

$$p(x,t) = p_0 + \frac{\Delta p}{2(\pi D_p t)^{0.5}} e^{-\frac{(x-vt)^2}{4D_p t}}, \quad (12.24)$$

where Δp is the number of holes per unit area created in a pulse spread over a negligibly small distance at $t = 0$. (Hint: try substituting the expression into the equations and seeing what happens.)

(12.3)* An n-type Ge sample is used in a Shockley–Haynes experiment. Minority carriers are injected using a flash of light. For a distance of 10 mm between the light spot and the detector contact, and in an electric field of 200 Vm^{-1}, a pulse of width 117 μs arrives at the contact 250 μs after the light has been pulsed. Using eqn 12.24 or otherwise, *estimate* the hole mobility and the diffusion coefficient.

(12.4) Why is it necessary to use either rectifying contacts or a pulse of light to inject the carriers in a Shockley–Haynes experiment? Why is it impossible to measure the majority carrier mobility in this way? What is the relevance of the Shockley–Haynes experiment to the design and performance of junction transistors?

(12.5) A current flows through a semiconductor pn junction. Describe, with the aid of sketches, the relative importance of majority and minority carrier **drift** and **diffusion** currents in the various regions of the device, and derive an expression relating the current through the diode to the bias voltage.

(12.6) Consider a distribution of free (mobile) charge, of density ρ_{free} Cm^{-3}. By considering the flux of charge out of a volume (or otherwise) show that the charge must obey the continuity equation

$$-\frac{\partial \rho_{\text{free}}}{\partial t} = \nabla \cdot \mathbf{J},$$

where $\mathbf{J}$ is the associated current density. Now $\mathbf{J} = \sigma \mathbf{E}$, where σ is the conductivity; using this fact and Maxwell's first equation ($\nabla \cdot \mathbf{D} = \rho_{\text{free}}$), show that

$$\rho_{\text{free}} = \rho_{\text{free}0} e^{-(t/\tau')},$$

where $\tau' = \epsilon_r \epsilon_0 / \sigma$.

A crystal of germanium has $\epsilon_r \approx 16$ and $\sigma \approx 10 \ \Omega^{-1} \text{m}^{-1}$. Evaluate τ', and hence show that any abnormal distribution of majority carrier density disappears in a time ~ 10 ps.

Estimate τ' for a typical metal (see Chapter 1 for typical metallic conductivities).

(12.7) A quantum point contact device formed by evaporating narrow metal gates on top of a GaAs-(Ga,Al)As heterojunction (see Fig. 12.6). Describe a method for fabricating such a heterojunction, indicating how **modulation doping** could be used to ensure that the electrons in the two-dimensional electron gas (2DEG) at the interface have a very high mobility. Explain how the energy of the conduction band-edge in the heterojunction varies as a function of position in the direction perpendicular to the 2DEG, accompanying your description with a sketch.

Estimate the typical spacing of the one-dimensional subbands in the one-dimensional constriction under the gates, given that the gap between the gates is 0.3 μm. You may assume that the effective mass of the electrons in the 2DEG is $0.07m_e$.

Describe the conductance of the quantum point contact as a function of gate voltage. Suggest an upper limit for the temperature at which conductance plateaux could be observed.

Useful terminology in condensed matter physics

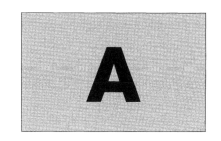

A.1 Introduction

In this book we deal exclusively with crystalline materials. It is therefore important to define some of the more commonly-encountered terms in crystallography.

A.2 Crystal

A crystal consists of a repeating pattern of objects (e.g. atoms or molecules) in an effectively infinite three-dimensional array. A convenient way of describing a crystal employs a lattice combined in some way with a physical object, sometimes called by physicists a basis.

A.3 Lattice

A lattice is an infinite array of points in space, in which each point has identical surroundings to all other points. A convenient way of representing a lattice is to use the the points generated by a translation vector

$$\mathbf{T} = n_1\mathbf{a}_1 + n_2\mathbf{a}_2 + n_3\mathbf{a}_3, \qquad (A.1)$$

where the $\mathbf{a}_j$s are three non-coplanar vectors and the n_js are numbers which repeat in a regular fashion.[1] If the $\mathbf{a}_j$ are such that all of the points on the lattice can be reached using *integer* n_j, then the $\mathbf{a}_j$ are said to be primitive.

A.4 Basis

A basis is an atom, ion, or collection of atoms or ions (e.g. a molecule). We form a crystal by combining the basis with the lattice in an operation known as *convolution*; a copy of the basis is attached to each point defined by the vectors $\mathbf{T}$. This is illustrated by Fig. A.1. In the simplest cases, the basis may consist of a single atom, so that the resulting crystal structure looks rather like the picture of the lattice. However, the two are *not* the same; the lattice consists of *points*, not atoms.

A.1	Introduction	165
A.2	Crystal	165
A.3	Lattice	165
A.4	Basis	165
A.5	Physical properties of crystals	166
A.6	Unit cell	166
A.7	Wigner–Seitz cell	167
A.8	Designation of directions	167
A.9	Designation of planes; Miller indices	168
A.10	Conventional or primitive?	169
A.11	The 14 Bravais lattices	171

[1] For example $n_j = 0, \frac{1}{3}, 1, 1\frac{1}{3}, 2, 2\frac{1}{3}, 3, 3\frac{1}{3}\ldots$.

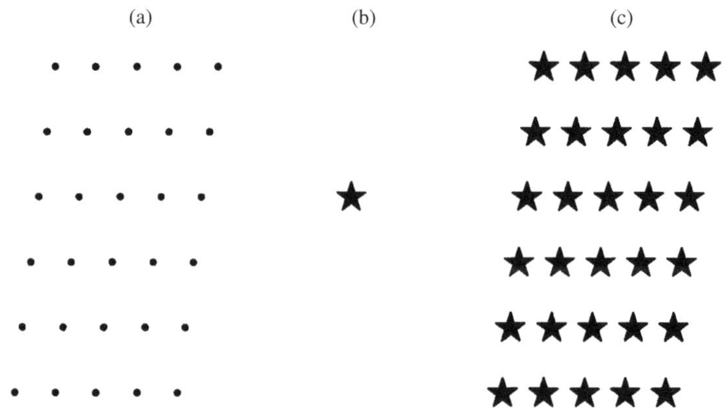

Fig. A.1 Illustration of the terms lattice, basis and crystal; for simplicity, a two-dimensional model is used. (a) Two-dimensional lattice of points generated by the vectors $\mathbf{T} = n_1\mathbf{a}_1 + n_2\mathbf{a}_2$ (see eqn A.1). (b) A basis (star). (c) A 'crystal' formed by convolving (a) with (b) (i.e. attaching a basis to each lattice point).

A.5 Physical properties of crystals

Having attached our bases (atoms or molecules) to the lattice, we can formulate another useful definition involving the lattice; the arrangement of atoms will look exactly the same in every respect at two points $\mathbf{r}$ and $\mathbf{r}'$ if

$$\mathbf{r} = \mathbf{r}' + \mathbf{T},$$

where $\mathbf{T}$ is defined as in eqn A.1. Any property of the system will also exhibit this periodicity; for example, if the electron potential energy is $V(\mathbf{r})$ at a point $\mathbf{r}$, then it should be the same at a point $\mathbf{r} + \mathbf{T}$, i.e.

$$V(\mathbf{r}) = V(\mathbf{r} + \mathbf{T}).$$

This leads to another definition of the term 'primitive', completely equivalent to that given in Section A.3; the lattice and the vectors $\mathbf{a}_j$ are said to be **primitive** if any two points at which the atomic arrangement looks the same can be connected by a vector $\mathbf{T}$ using a suitable choice of integer n_j. This is illustrated for the body-centred cubic and face-centred cubic lattices in Figs. A.2 and A.3.

A.6 Unit cell

A unit cell is a region of space that when repeated fills all space. In a crystal structure, there is an infinite set of choices of unit cell, three of which are illustrated in Fig. A.4 for the simple 'crystal' in Fig. A.1. Unit cells A and C are minimum volume unit cells for this structure; that is, they contain one basis (or one lattice point) each. This can be easily seen if one visualises moving cell A or C along its body diagonal slightly; three lattice points are left outside and one is wholly within. Alternatively, one can say that each of the bases touched by cell A or C is shared by four other cells; in other words, unit cell A or C 'contains' $4 \times \frac{1}{4} = 1$ basis. Such minimum volume unit cells are known as *primitive unit cells*. In three dimensions a primitive unit cell will have the volume $\mathbf{a}_1.(\mathbf{a}_2 \times \mathbf{a}_3)$.

Unit cell B has twice the volume of A or C; it is known as a centred unit cell.

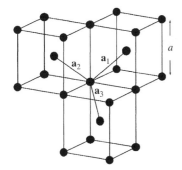
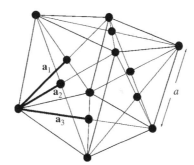
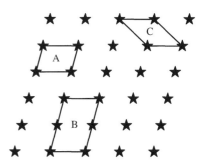

Fig. A.2 The body-centred cubic lattice (also known as the cubic-I lattice); the lattice points are represented by dots. At first sight, it might seem that a suitable choice of lattice translation vectors would be based upon the sides of the cubes, of length a. If the cube edges are assumed to be parallel to the Cartesian unit vectors, this would lead to $\mathbf{T} = n_1 a \mathbf{e}_1 + n_2 a \mathbf{e}_2 + n_3 a \mathbf{e}_3$. However, such a set of vectors would *not* permit translation from a point on the cube corner to a point at the cube centre using integer n_j; in other words, they are not primitive lattice translation vectors. Instead, the primitive lattice translation vectors are based upon $\mathbf{a}_1 = \frac{a}{2}(-\mathbf{e}_1 + \mathbf{e}_2 + \mathbf{e}_3)$, $\mathbf{a}_2 = \frac{a}{2}(\mathbf{e}_1 - \mathbf{e}_2 + \mathbf{e}_3)$ and $\mathbf{a}_3 = \frac{a}{2}(\mathbf{e}_1 + \mathbf{e}_2 - \mathbf{e}_3)$ (shown in the figure). Using these vectors, a translation from any lattice point to any other lattice point can be accomplished by $\mathbf{T} = n_1 \mathbf{a}_1 + n_2 \mathbf{a}_2 + n_3 \mathbf{a}_3$ using suitable n_j. (Note that the cubes represent conventional unit cells; see Section A.10.)

Fig. A.3 The face-centred cubic lattice (also known as the *cubic-F lattice*); the lattice points are represented by dots. As in Fig. A.2, it at first looks as though a suitable choice of lattice translation vectors would be based upon the sides of the cubes, again of length a. However, we run into a similar problem to that described in the caption to Fig. A.2; such a set of vectors does not permit one to translate from a point on the cube corner to a point on the cube face. Instead, the primitive lattice translation vectors are based upon $\mathbf{a}_1 = \frac{a}{2}(\mathbf{e}_2 + \mathbf{e}_3)$, $\mathbf{a}_2 = \frac{a}{2}(\mathbf{e}_1 + \mathbf{e}_3)$ and $\mathbf{a}_3 = \frac{a}{2}(\mathbf{e}_1 + \mathbf{e}_2)$ (shown in the figure). Using these vectors, a translation from any lattice point to any other lattice point can be accomplished by $\mathbf{T} = n_1 \mathbf{a}_1 + n_2 \mathbf{a}_2 + n_3 \mathbf{a}_3$ using suitable n_j. (Note that the cube represents the conventional unit cell; see Section A.10.)

Fig. A.4 Three possible unit cells for the two-dimensional crystal of Fig. A.1. A and C are two minimum volume unit cells, each containing one basis. B has twice the volume of A or C.

A.7 Wigner–Seitz cell

There are obviously many ways of choosing a primitive unit cell. It is very convenient to choose a primitive unit cell which is centred about a lattice point, and which therefore explicitly reflects the symmetry of the underlying lattice. Such a cell is called a **Wigner–Seitz** cell. The cell is obtained by starting at any lattice point (which is then the origin) and constructing vectors to all neighbouring lattice points. Planes are then constructed perpendicular to and passing through the midpoints of these vectors. The Wigner–Seitz cell is the cell with the smallest volume about the origin bounded by these planes.[2] Figure A.5 illustrates the construction of Wigner–Seitz cells for two different lattices.

A.8 Designation of directions

Directions in crystals are designated by the following process;

(1) a lattice vector $\mathbf{T} = n'_1 \mathbf{a}_1 + n'_2 \mathbf{a}_2 + n'_3 \mathbf{a}_3$, where n'_1, n'_2 and n'_3 are integers, is found which is parallel to the chosen direction;

[2]We shall come to define the reciprocal lattice in Chapter 2. The first Brillouin zone is usually taken to be the Wigner–Seitz cell of the Reciprocal lattice; as the reciprocal lattice point is at the centre, the Wigner–Seitz cell displays the full symmetry of the reciprocal lattice. This is an excellent idea in crystals of high symmetry, where the degeneracies predicted on the grounds of symmetry can be visualised by merely looking at the Brillouin zone. However, in some low-symmetry crystals, the Wigner–Seitz construction leads to a rather complex-looking unit cell which can obscure the physics. In such cases a simple parallelepiped may be a better choice of unit cell.

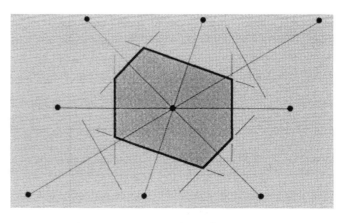
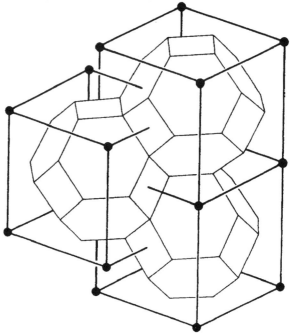

Fig. A.5 Construction of the Wigner–Seitz cell. The top part of the figure shows the construction of the Wigner–Seitz cell for a simple two-dimensional lattice. Vectors are constructed from a lattice point chosen as the origin to all neighbouring lattice points. Planes (lines in two dimensions) are then constructed perpendicular to and passing through the midpoints of these vectors. The Wigner–Seitz cell is the cell with the smallest volume about the origin bounded by these planes; in the figure it is shaded a darker grey. The lower part shows the stacking of Wigner–Seitz cells in a body-centred cubic lattice; lattice points within the cells are not shown.

(2) the integers n'_1, n'_2 and n'_3 are reduced to the set of simplest integers n_1, n_2 and n_3 by dividing by their highest common factor;

(3) the direction is then labelled as $[n_1 n_2 n_3]$.

If a negative value of one of the integers was required to describe the direction, then this is denoted by a bar, e.g. $\bar{n}_1$. The complete set of equivalent directions in a crystal is denoted by using angular brackets, i.e. $< n_1 n_2 n_3 >$.[3]

[3] It is customary in crystallography to use $[uvw]$ (rather than $[n_1 n_2 n_3]$) as the algebraic symbols denoting directions.

A.9 Designation of planes; Miller indices

Miller indices are integers m_1, m_2, m_3 that refer to the intercepts made by a plane on axes defined by $\mathbf{a}_1$, $\mathbf{a}_2$ and $\mathbf{a}_3$ (see Figure A.6). An $(m_1 m_2 m_3)$ plane

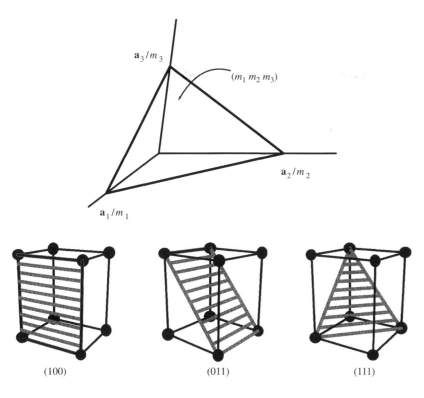

Fig. A.6 Definition of Miller indices; an $(m_1 m_2 m_3)$ plane cuts the $\mathbf{a}_1$ axis at $\mathbf{a}_1/m_1$, the $\mathbf{a}_2$ axis at $\mathbf{a}_2/m_2$ and the $\mathbf{a}_3$ axis at $\mathbf{a}_3/m_3$.

Fig. A.7 The use of Miller indices in defining the most important planes of the simple cubic lattice. The cubes in the diagrams indicate lattice points in the simple cubic system; the 'origin' is taken to be the lower rearmost point of each cube. The primitive lattice translation vectors for the simple cubic lattice are based on $\mathbf{a}_1 = a\mathbf{e}_1$, $\mathbf{a}_2 = a\mathbf{e}_2$, $\mathbf{a}_3 = a\mathbf{e}_3$, where a is the length of the cube edge. Starting on the left, the plane cuts the $\mathbf{a}_1$ axis at $\mathbf{a}_1$, the $\mathbf{a}_2$ axis at infinity and the $\mathbf{a}_3$ axis at infinity; i.e. $1/m_1 = 1$, $1/m_2 = 1/m_3 = \infty$. The plane therefore has the Miller indices (100). In the centre figure, the plane cuts the $\mathbf{a}_1$ axis at infinity, the $\mathbf{a}_2$ axis at $\mathbf{a}_2$ and the $\mathbf{a}_3$ axis at $\mathbf{a}_3$; i.e. $1/m_1 = \infty$, $1/m_2 = 1/m_3 = 1$. The plane therefore has the Miller indices (011). The right-hand figure shows the (111) plane, deduced using similar reasoning.

cuts the $\mathbf{a}_1$ axis at $\mathbf{a}_1/m_1$, the $\mathbf{a}_2$ axis at $\mathbf{a}_2/m_2$ and the $\mathbf{a}_3$ axis at $\mathbf{a}_3/m_3$.[4] As before, a bar is used for a negative intercept. The general set of planes of a particular type is labelled using curly brackets, i.e. $\{m_1 m_2 m_3\}$.

Figure A.7 demonstrates the use of Miller indices in defining some important planes in the simple cubic lattice.

[4] It is customary in crystallography to use (hkl) (rather than $(m_1 m_2 m_3)$) as the algebraic symbols representing Miller indices.

A.10 Conventional or primitive?

Our primary interest in this book is the electrons which inhabit the crystal. We shall see that in many respects the behaviour of these electrons is determined by the underlying symmetry of the lattice. Therefore, as far as the electrons are concerned, the lattice should always be described by the primitive lattice translation vectors; these encompass the basic, underlying periodicity of the crystal. Likewise, we shall see that when the filling of bands by electrons is considered, it is the number of electrons provided by each **primitive unit cell** of the crystal that is important.

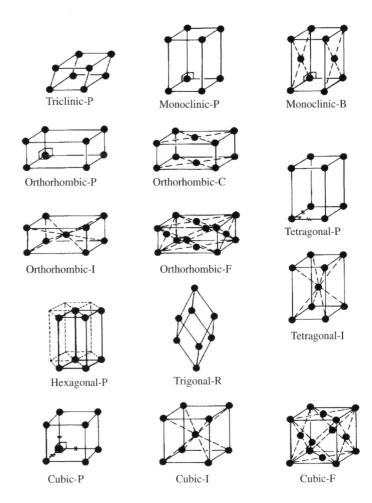

Fig. A.8 The conventional unit cells of the 14 Bravais lattices.

Unfortunately, many human beings prefer to think in Cartesian coordinates, and so it is often customary to use so-called *conventional* unit cells and lattice vectors. These are designed so that as many of the chosen $\mathbf{a}_j$ are perpendicular to each other as possible. This almost inevitably results in a lattice vectors and unit cells which are non-primitive. In Figs. A.2 and A.3, the conventional unit cells of the body-centred cubic and face-centred cubic lattices are shown as cubes, with the conventional lattice vectors forming the edges of the cubes. The conventional unit cell of the body-centred cubic lattice thus contains two lattice points, and that of the face-centred cubic lattice four lattice points. In the following section, all the possible three-dimensional conventional unit cells are shown in Fig.A.8.

In summary, you will find that data concerning crystals (directions within a crystal, Miller indices etc.) will almost always be quoted using conventional coordinates rather than primitive ones (see e.g. Exercise 1.1). However, as far as electrons are concerned, it is the primitive coordinates of the lattice that are important.[5] When thinking about the behaviour of electrons, you should always try to visualise the situation in terms of the primitive (usually Wigner–Seitz) cell in r-space and the first Brillouin zone in k-space.

[5]The reason for this difference is that in the case of electrons (or phonons) we are interested only in counting states, not their real-space positions. In crystallography, we are interested in the atomic positions; these are usually more understandable with higher symmetry cells, even if this means using a larger cell than necessary.

A.11 The 14 Bravais lattices

Thus far, we have only specified the translational invariance of the lattice, by defining the translation vectors **T**. However, a lattice may also be invariant under symmetry operations such as rotation through a certain angle, inversion and reflection. If these properties are used to classify a lattice, it turns out that there are only 14 distinct types of lattice in three dimensions that can fill all space. These are referred to as the 14 space lattices or the 14 Bravais lattices. Figure A.8 shows the conventional unit cells of all 14 Bravais lattices.

Further reading

As this book's primary purpose is to describe electronic bandstructure, this Appendix has presented only a brief summary. More extended treatments may be found in *Space groups for solid state scientists*, by Gerald Burns and A.M. Glazer (Academic Press, New York, 1978), Chapters 2 to 4 (especially recommended), *Introduction to solid state physics*, by Charles Kittel, seventh edition (Wiley, New York, 1996) Chapters 1 and 2, *The solid state*, by H.M. Rosenberg (OUP, Oxford, 1988), Chapter 1 and in *Solid state physics*, by G. Burns (Academic Press, Boston, 1995).

B Derivation of density of states in k-space

B.1 Introduction

It is very important to be able to enumerate the number of states available to particles within a solid. We consider free particles moving within (confined by) a rectangular parallelepiped of dimensions L_x, L_y and L_z in the x, y and z directions respectively. It is possible to derive the number of quantum-mechanical states within such a box by using wavefunctions which are standing waves. However, in the present context we are interested in transport phenomena (electrical conductivity etc.) which implies that travelling waves are more useful. We therefore apply **periodic boundary conditions**,

$$\psi(x, y, z) = \psi(x + L_x, y, z) = \psi(x, y + L_y, z) = \psi(x, y, z + L_z). \quad (B.1)$$

The solutions to eqn B.1 are plane waves

$$\psi = a e^{i\mathbf{k}\cdot\mathbf{r}}, \quad (B.2)$$

where a is a normalising constant. The boundary conditions B.1 constrain $\mathbf{k}=(k_x, k_y, k_z)$ to take values

$$k_x = \pm \frac{2n_x\pi}{L_x}, \quad k_y = \pm \frac{2n_y\pi}{L_y}, \quad k_z = \pm \frac{2n_z\pi}{L_z},$$

where n_x, n_y and n_z are integers (as $e^{2\pi i j} = 1$ when j is an integer); we can map out these states as a grid of points in k-space (see Fig. B.1). Thus, the wavevector is quantised, and each time we increase n_x, n_y or n_z by one, we obtain a new state; hence, the volume of k-space per state is $(2\pi)^3/(L_x L_y L_z)$. The number of states per unit volume of k-space is the reciprocal of this; therefore, the number of states N contained within a three dimensional k-space volume V_{k3} is

$$N = \frac{1}{(2\pi)^3} V_{k3} V_{r3}, \quad (B.3)$$

where $V_{r3} = L_x L_y L_z$ is the r-space volume of the system. It is simple to show that a similar relationship applies in two dimensions, i.e.,

$$N = \frac{1}{(2\pi)^2} V_{k2} V_{r2}, \quad (B.4)$$

where $V_{r2} = L_x L_y$ is the r-space area of the system and V_{k2} is the k-space area. In fact, it should be apparent that in a general number of dimensions j

$$N = \frac{1}{(2\pi)^j} V_{kj} V_{rj}, \quad (B.5)$$

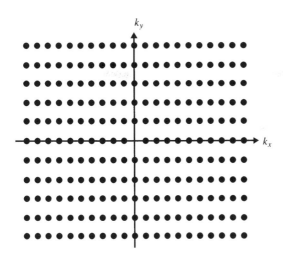

Fig. B.1 A map of states in k-space. The centre of the graph is at $k_x = k_y = 0$.

where V_{rj} and V_{kj} are generalised 'volumes' (lengths, areas, volumes) of r and k-space.

Note that it is possible to derive eqns B.3, B.4 and B.5 using a variety of different 'box' shapes and boundary conditions; the end result is always the same (try it) and independent of the r-space 'volume' used.

B.1.1 Density of states

We now consider a volume of three-dimensional k-space which contains states with wavevectors of magnitude $|\mathbf{k}|$ or smaller; this will comprise a sphere, radius $k = |\mathbf{k}|$, leading to $V_{k3} = \frac{4}{3}\pi k^3$. We insert this in eqn B.3, dividing by V_{r3} to obtain

$$n = \frac{N}{V_{r3}} = \frac{1}{(2\pi)^3}\frac{4}{3}\pi k^3 = \frac{k^3}{6\pi^2}, \quad (B.6)$$

where we have defined n, the number of states per unit volume of r-space. The density of states, $g(k)$ is defined as follows: the number of states per unit volume of r-space with wavevector between k and $k + \mathrm{d}k$ is $g(k)\mathrm{d}k$. From this, it should be obvious that

$$g(k) \equiv \frac{\mathrm{d}n}{\mathrm{d}k} = \frac{k^2}{2\pi^2}. \quad (B.7)$$

Up to this point, the derivation has not specified *what* the particles are; all we have done is derived the number of k-states per unit volume per interval of k. Thus, *the derivation is universal and independent of the particle considered.* It will work for phonons, electrons, muons, Higgs bosons or anything that one might care to consider.

Sometimes it may be convenient to deal with a $g(k)$ which is defined in terms numbers of single-particle states, rather than numbers of k-states. For example, electrons are fermions; Pauli's exclusion principle only permits one electron per quantum state. However, the electron has spin $S = \frac{1}{2}$, and therefore two possible values of S_z, $\pm\frac{1}{2}$. Each k state will therefore be able to accommodate two electrons, one with $S_z = +\frac{1}{2}$ and one with $S_z = -\frac{1}{2}$. For **electron single-**

particle states, eqn B.7 therefore becomes

$$g(k) = 2\frac{k^2}{2\pi^2} = \frac{k^2}{\pi^2}. \tag{B.8}$$

The spin degeneracy of the electrons has merely multiplied eqn B.7 by a numerical factor. We can therefore *always* say that in three dimensions, whatever the particle, and independent of whether $g(k)$ is defined in terms of numbers of k states or numbers of single-particle states

$$g(k) \propto k^2. \tag{B.9}$$

Considering eqns B.5 and B.6, it will be observed that in j dimensions, $n \propto k^j$. Therefore, remembering that $g(k) \equiv \mathrm{d}n/\mathrm{d}k$, we can also be quite confident that in j dimensions

$$g(k) \propto k^{j-1}. \tag{B.10}$$

We shall see that the differing forms of these relationships will have a fundamental effect on the behaviour of systems of different dimensionality.

B.1.2 Reading

There are treatments of this topic in almost every quantum mechanics book; see e.g. *Quantum mechanics*, by Stephen Gasiorowicz (Wiley, New York, 1974), page 348.

Derivation of distribution functions

C.1 Introduction

I have invoked the distribution functions for various types of particle and excitation in a number of places in this book, and so I thought it prudent to include a short derivation, which shows how the distribution functions reflect the underlying quantum-mechanical behaviour of the particles or excitations. First of all, I shall provide a very quick derivation of the **grand partition function**, which predicts the behaviour of a collection of particles (the *system*) in thermal equilibrium with a large number of similar collections of particles (the **heat bath**). Then I shall show how the grand partition function factorises in the case of non-interacting particles, leading to the **distribution functions**, which describe the occupation of single-particle energy levels.

We consider a system inside a heat bath, which consists of a very large number of systems which are similar to the system under study. The system contains a variable number of particles N, has a variable energy E and occupies a fixed volume V; it is able to exchange energy and particles with the surrounding heat bath. The combination of heat bath and system has energy E_0, and volume V_0 and contains N_0 particles; it is isolated from its surroundings, i.e. it is unable to exchange particles or energy with its surroundings.

For a certain value of N, let the system have states with energies

$$E_{N1} \leq E_{N2} \leq E_{N3} \leq E_{N4} ..... \leq E_{Nj} \leq ...;$$

here, N and the index j serve as a label for the energy state of the system. When the system is in the state labelled E_{Nj}, then the heat bath will have energy $E_0 - E_{Nj}$, contain $N_0 - N$ particles and occupy a volume $V_0 - V$.

At this point, we come to some of the key assumptions of statistical physics. The state of the bath and system has been defined using 'macroscopic' measurable variables (energy E, volume V, particle number N (the latter could be measured macroscopically by weighing the system)); such a definition is known as a *macrostate*. However, if the system and bath comprise many particles, there will be very many ways of arranging the energy states of the particles to produce the macrostate. Each one of these descriptions is known as a **microstate**; it is a quantum state of the whole assembly (bath plus system). The most fundamental assumption of statistical mechanics is the **averaging postulate**, i.e. all accessible microstates are equally probable. This implies that the probability of finding a distinct macrostate is simply proportional to the number of microstates which correspond to that macrostate; this number of microstates is known as the **statistical weight** Ω of the macrostate.

Let the statistical weight of the macrostate of the heat bath be $\Omega = \Omega(E_0 - E_{Nj}, V_0 - V, N_0 - N)$ when the system is in the state Nj. The probability p_{Nj}

of finding the system in the state with energy E_{Nj} will be proportional to Ω, i.e.

$$p_{Nj} = G \times \Omega(E_0 - E_{Nj}, V_0 - V, N_0 - N), \tag{C.1}$$

where G is a normalisation constant. Boltzmann's definition of the entropy S of the heat bath is

$$S = k_B \ln \Omega. \tag{C.2}$$

Combining these two equations leads to

$$p_{Nj} = G e^{S/k_B}. \tag{C.3}$$

If the system is very small compared to the heat bath, then $V \ll V_0$, $N \ll N_0$ and $E_{Nj} \ll E_0$. Under such circumstances, we can use Taylor's theorem:

$$S(E_0 + \Delta E_0, V_0 + \Delta V_0, N_0 + \Delta N_0) = S(E_0, V_0, N_0) +$$

$$\Delta E_0 \left(\frac{\partial S}{\partial E_0}\right)_{V_0, N_0} + \Delta V_0 \left(\frac{\partial S}{\partial V_0}\right)_{E_0, N_0} + \Delta N_0 \left(\frac{\partial S}{\partial N_0}\right)_{E_0, V_0},$$

where ΔE_0, ΔN_0 and ΔV_0 are small variations in E_0, N_0 and V_0 respectively. The entropy of the bath is

$$S(E_0 - E_{Nj}, V_0 - V, N_0 - N) = S(E_0, V_0, N_0) -$$

$$E_{Nj} \left(\frac{\partial S}{\partial E_0}\right)_{V_0, N_0} - V \left(\frac{\partial S}{\partial V_0}\right)_{E_0, N_0} - N \left(\frac{\partial S}{\partial N_0}\right)_{E_0, V_0}. \tag{C.4}$$

One of the thermodynamic definitions of temperature T is

$$\frac{1}{T} = \left(\frac{\partial S}{\partial E_0}\right)_{N_0, V_0}. \tag{C.5}$$

In a similar way, the chemical potential, μ, is defined as

$$\mu = -T \left(\frac{\partial S}{\partial N_0}\right)_{E_0, V_0}. \tag{C.6}$$

Thus

$$S = S(E_0, V_0, N_0) - V \left(\frac{\partial S}{\partial V_0}\right)_{E_0, N_0} - \frac{E_{Nj}}{T} + \frac{\mu N}{T}; \tag{C.7}$$

remembering that V is a constant, we can substitute eqn C.7 into eqn C.3 to obtain

$$p_{Nj} = \frac{e^{(\mu N - E_{Nj})/k_B T}}{\mathcal{Z}}, \tag{C.8}$$

where we have normalised by dividing by

$$\mathcal{Z} = \sum_{N=0}^{N_0} \left(\sum_{j=1}^{\infty} e^{(\mu N - E_{Nj})/k_B T} \right), \tag{C.9}$$

as the total probability of all possible Nj must be one. $\mathcal{Z}$ is usually known as the **grand partition function**. Equation C.8 gives the probability that the system occupies the particular macrostate with energy E_{Nj}.

We now specify the state of the system in more detail by defining a set of single-particle states (i.e. states of particles which do not interact) with energies $\mathcal{E}_1 \leq \mathcal{E}_2 \leq \mathcal{E}_3.... \leq \mathcal{E}_l....$ These states are occupied by $n_1, n_2, n_3.......n_l...$ particles respectively. Thus the energy of the system is given by $E_{Nj} = \sum_l n_l \mathcal{E}_l$ and the number of particles in it by $N = \sum_l n_l$. Substituting this into eqn C.8, yields

$$p_{Nj} = \frac{e^{\{\mu(n_1+n_2+n_3...)-(n_1\mathcal{E}_1+n_2\mathcal{E}_2+n_3\mathcal{E}_3......)\}/k_BT}}{\mathcal{Z}}, \tag{C.10}$$

which factorises into

$$p_{Nj} = \frac{\{e^{(\mu-\mathcal{E}_1)n_1/k_BT}\}\{e^{(\mu-\mathcal{E}_2)n_2/k_BT}\}\{e^{(\mu-\mathcal{E}_3)n_3/k_BT}\}.....}{\mathcal{Z}}. \tag{C.11}$$

Similarly, the grand partition function becomes

$$\mathcal{Z} = \sum_{n_1,n_2,n_3...} e^{(\mu(n_1+n_2+....)-(n_1\mathcal{E}_1+n_2\mathcal{E}_2+n_3\mathcal{E}_3...))/k_BT}, \tag{C.12}$$

where each occupation number n_l is summed over all possible values; it is relatively easy to see that this expression also factorises;

$$\mathcal{Z} = \sum_{n_1,n_2,n_3...} \{e^{(\mu-\mathcal{E}_1)n_1/k_BT}\}\{e^{(\mu-\mathcal{E}_2)n_2/k_BT}\}\{e^{(\mu-\mathcal{E}_3)n_3/k_BT}\}..., \tag{C.13}$$

or, equivalently

$$\mathcal{Z} = \{\sum_{n_1} e^{(\mu-\mathcal{E}_1)n_1/k_BT}\}\{\sum_{n_2} e^{(\mu-\mathcal{E}_2)n_2/k_BT}\}\{\sum_{n_3} e^{(\mu-\mathcal{E}_3)n_3/k_BT}\}... \tag{C.14}$$

Combining eqns C.11 and C.14, yields

$$p_{Nj} = \prod_{l=1}^{\infty} p_l(n_l), \tag{C.15}$$

where

$$p_l(n_l) = \frac{e^{(\mu-\mathcal{E}_l)n_l/k_BT}}{\mathcal{Z}_l} \quad \text{and} \quad \mathcal{Z}_l = \sum_{n_l} e^{(\mu-\mathcal{E}_l)n_l/k_BT}. \tag{C.16}$$

The factorisation of the equation for p_{Nj} means that the probability p_l of finding n_l particles in the lth single-particle state (that with energy $\mathcal{E}_l$) is independent of the occupancies of the other states. This has occurred because we specified that the particles be non-interacting; i.e. the occupation of a particular energy state does not influence the energies or occupancies of the others. To treat the system, it is therefore sufficient to deal with the probability of occupation of each single-particle state l, as defined by eqns C.16.

Before going further, we note that the mean occupation number $< n_l >$ of the lth single-particle state is

$$< n_l > = \sum_{n_l=0}^{\infty} n_l p_l(n_l). \tag{C.17}$$

Inspection of eqns C.16 shows that this can also be written as

$$< n_l > = k_BT \left(\frac{\partial \ln \mathcal{Z}_l}{\partial \mu} \right)_{V,T}. \tag{C.18}$$

We now introduce some constraints on the numbers n_l.

C.1.1 Bosons

Bosons are particles or excitations with integer (0, 1, 2...) spin. In the case of bosons, n_l is unconstrained, i.e. $n_l = 0, 1, 2, 3......$ The expression for $\mathcal{Z}_l$ becomes

$$\mathcal{Z}_l = \sum_{n_l=0}^{\infty} e^{(\mu-\mathcal{E}_l)n_l/k_B T} = \left(1 - e^{(\mu-\mathcal{E}_l)/k_B T}\right)^{-1}. \tag{C.19}$$

Equation C.18 therefore yields

$$<n_l> = k_B T \left(-\frac{1}{k_B T}\right) \frac{e^{(\mu-\mathcal{E}_l)/k_B T}}{e^{(\mu-\mathcal{E}_l)/k_B T} - 1},$$

which simplifies to

$$<n_l> = \frac{1}{e^{(\mathcal{E}_l-\mu)/k_B T} - 1}. \tag{C.20}$$

This is the **Bose–Einstein** distribution function.

C.1.2 Fermions

Fermions are particles with half-integer spin ($\frac{1}{2}, \frac{3}{2}$...). Pauli's exclusion principle states that each single-particle state may be occupied by at most one fermion, limiting n_l to 0 or 1. Thus, using eqns C.16

$$\mathcal{Z}_l = 1 + e^{(\mu-\mathcal{E}_l)/k_B T}. \tag{C.21}$$

Using eqn C.18 yields

$$<n_l> = \frac{1}{e^{(\mathcal{E}_l-\mu)/k_B T} + 1}. \tag{C.22}$$

This is the *Fermi–Dirac distribution function*.

C.1.3 The Maxwell–Boltzmann distribution function

The Maxwell–Boltzmann distribution is the distribution function for classical particles; I include it here because it is important in showing why the Drude model failed. Classical particles are **distinguishable**, in contrast to quantum-mechanical particles, which are not. We must therefore treat each classical partical individually; we define a single-particle partition function

$$\mathcal{Z}_s = \sum_l e^{-\mathcal{E}_l/k_B T}. \tag{C.23}$$

This is to be evaluated for each particle which can inhabit the energy levels $\mathcal{E}_l$.

To assist in deriving the Maxwell–Boltzmann distribution, we factorise eqn C.9, to give

$$\mathcal{Z} = \sum_N e^{\mu/k_B T} \sum_j e^{-E_{Nj}/k_B T}. \tag{C.24}$$

For a gas of classical (distinguishable) particles,

$$\sum_j e^{-E_{Nj}/k_B T} = \frac{\mathcal{Z}_s^N}{N!}. \tag{C.25}$$

There is insufficient space to give a detailed derivation of this relationship, which is covered in all statistical physics texts; we merely remark that the $N!$ has come about because the particles are distinguishable, i.e. it matters which particle goes into which energy level. Hence, $\mathcal{Z}$ becomes

$$\mathcal{Z} = \sum_N \frac{(e^{\mu/k_BT}\mathcal{Z}_s)^N}{N!} = e^{\mathcal{Z}_s e^{\mu/k_BT}}, \tag{C.26}$$

where the final step has use the standard series expansion for an exponential. The mean value of N, $<N>$, may be evaluated by the following step:

$$<N> = k_BT\left(\frac{\partial \mathcal{Z}}{\partial \mu}\right)_{V,T} = \mathcal{Z}_s e^{\mu/k_BT}. \tag{C.27}$$

Now $<N>$ is also equivalent to the sum of the mean occupancies $<n_l>$ of all of the single-particle levels. Hence, comparing this with eqn C.27, we can write

$$<N> = \sum_l <n_l> = \sum_l e^{(\mu-\mathcal{E}_l)/k_BT}, \tag{C.28}$$

leading to the Maxwell–Boltzmann distribution function

$$<n_l> = e^{(\mu-\mathcal{E}_l)/k_BT}. \tag{C.29}$$

C.1.4 Mean energy and heat capacity of the classical gas

In the previous appendix, we saw that $g(k)$, the density of states in k space, was proportional to k^{j-1}, where j is the number of dimensions. We shall now evaluate the mean energy of a free particle obeying Maxwell–Boltzmann statistics. Given that the energy of a free particle of mass m is $\mathcal{E}(k) = \hbar^2 k^2/2m$, the quickest route is to evaluate the mean value of k^2, which we will call $<k^2>$; we do this by multiplying each value of k^2 by the appropriate Maxwell–Boltzmann factor (eqn C.29) and normalising appropriately, i.e.

$$<k^2> = \frac{\sum_k k^2 s(k) e^{(\mu-\mathcal{E}(k))/k_BT}}{\sum_k s(k) e^{(\mu-\mathcal{E}(k))/k_BT}}, \tag{C.30}$$

where the sums are over values of k, which also labels the energy. The parameter $s(k)$ is the number of states possessing a particular value of k, the occupancy of each of which will be determined by eqn C.29. The denominator is there to ensure correct normalisation. We assume that the states are very close together in energy, so that the sums can be replaced by integrals, and $s(k)$ replaced by $g(k)$. Noting that e^{μ/k_BT} cancels in eqn C.30, we obtain

$$<k^2> = \frac{\int_0^\infty k^2 k^{j-1} e^{-\alpha k^2} dk}{\int_0^\infty k^{j-1} e^{-\alpha k^2} dk} = j\frac{mk_BT}{\hbar^2}, \tag{C.31}$$

where $\alpha = \hbar^2/2mk_BT$ and the integrals are evaluated using standard tables. This shows that the mean value of the kinetic energy of a classical particle in j dimensions is

$$<E> \equiv \frac{\hbar^2 <k^2>}{2m} = \frac{j}{2}k_BT. \tag{C.32}$$

Derivation of distribution functions

The heat capacity C is therefore

$$C \equiv n\frac{\mathrm{d}<E>}{\mathrm{d}T} = \frac{j}{2}nk_\mathrm{B}. \tag{C.33}$$

Phonons

D

D.1 Introduction

The concept of *phonons* arises from attempts to describe the vibrational modes of crystals. In order to understand how this comes about, it is quite instructive to consider the crystal as an array of masses (the atoms or ions) connected by bonds, which may be imagined to be little springs joining the masses to their neigbours. We shall examine a simple model along these lines in the following section, and it will soon become apparent that such systems possess *normal modes* which have characteristic angular frequencies $\omega_j(\mathbf{k})$ corresponding to wavevectors $\mathbf{k}$; note that more than one frequency may be possible for a given value of $\mathbf{k}$, necessitating the additional label j.[1]

[1] The index j will later be seen to label a particular phonon *branch*.

In order to specify the vibrational energy of a three-dimensional crystal, it is usually regarded as an assembly of independent oscillators whose frequencies are the same as those of the normal modes. The total energy of the crystal contains contributions $\Delta U_{\mathbf{k},j}$ from each normal mode which take discrete values

$$\Delta U_{\mathbf{k},j} = (l_{\mathbf{k},j} + \frac{1}{2})\hbar\omega_j(\mathbf{k}),$$

where $l_{\mathbf{k},j} = 0, 1, 2\ldots$ is an excitation number for the mode with energy $\hbar\omega_j(\mathbf{k})$.[2] The total energy of the crystal U is then just given by summing all of the contributions from the individual modes:

[2] The additional $\frac{1}{2}$ within the equation represents the zero-point energy of the system due to that mode.

$$U = \sum_{\mathbf{k},j}\Delta U_{\mathbf{k},j} = \sum_{\mathbf{k},j}(l_{\mathbf{k},j} + \frac{1}{2})\hbar\omega_j(\mathbf{k}).$$

This manner of describing the state of the crystal is rather cumbersome, especially when one starts to consider the way in which the vibrational modes of the crystal exchange energy and/or $\mathbf{k}$ with electrons. To surmount this problem, a particulate description of the modes is used; rather than considering a normal mode $\omega_j(\mathbf{k})$ with excitation number $l_{\mathbf{k},j}$, we say that $l_{\mathbf{k},j}$ *phonons* with energy $\hbar\omega_j(\mathbf{k})$ are present. The electrons in the crystal can then be said to exchange energy and $\mathbf{k}$ with the vibrational state of the crystal by scattering from, emitting or absorbing phonons.

For this reason, phonons are cited as a scatterer of electrons in this text book. In order to see why they act as they do, it is necessary to give a brief account of their properties. We shall do this by using two simple models. The first, in the next Section, allows the different types of phonon mode to be understood. The second is the Debye model, which provides a very convenient energy scale, the Debye temperature, for the phonons in a particular material and which allows reasonable estimates of the number of phonons at a particular temperature to be made.

D.2 A simple model

Virtually all of the relevant properties of phonons can be derived by considering a diatomic, linear (one-dimensional) chain, consisting of alternate masses M_1 and M_2 with equilibrium separation a, i.e. the unit cell is of length $2a$ (see Fig. D.1). We label the displacement of the jth atom x_j, and assume a harmonic restoring force (constant of proportionality $\mathcal{F}$) between nearest neighbours, giving the following equations of motion;

$$M_1 \frac{d^2 x_{2j}}{dt^2} = \mathcal{F}(x_{2j+1} + x_{2j-1} - 2x_{2j})$$

and

$$M_2 \frac{d^2 x_{2j+1}}{dt^2} = \mathcal{F}(x_{2j+2} + x_{2j} - 2x_{2j+1}). \quad \text{(D.1)}$$

To solve, substitute the trial solutions

$$x_{2j} = \mathcal{A} e^{-i(\omega t - 2jqa)} \quad \text{and} \quad x_{2j+1} = \mathcal{B} e^{-i(\omega t - (2j+1)qa)},$$

where $\mathcal{A}$ and $\mathcal{B}$ are constants, into eqns D.1, to yield

$$-\omega^2 \mathcal{A} M_1 = \mathcal{F}\mathcal{B}(e^{iqa} + e^{-iqa}) - 2\mathcal{F}\mathcal{A}$$

and

$$-\omega^2 \mathcal{B} M_2 = \mathcal{F}\mathcal{A}(e^{iqa} + e^{-iqa}) - 2\mathcal{F}\mathcal{B}$$

which may be rearranged to give

$$(2\mathcal{F} - \omega^2 M_1)\mathcal{A} = 2\mathcal{F}\mathcal{B} \cos qa \quad \text{(D.2)}$$

$$2\mathcal{F}\mathcal{A} \cos qa = (2\mathcal{F} - \omega^2 M_2)\mathcal{B}. \quad \text{(D.3)}$$

Equations D.2 and D.3 may be solved simultaneously to give

$$\omega^2 = \mathcal{F}\left\{\left(\frac{1}{M_1} + \frac{1}{M_2}\right) \pm \left(\left(\frac{1}{M_1} + \frac{1}{M_2}\right)^2 - \frac{4}{M_1 M_2} \sin^2 qa\right)^{\frac{1}{2}}\right\}. \quad \text{(D.4)}$$

Figure D.2 shows the dispersion relationships predicted by eqn D.4. Note the following:

- there are two distinct branches, the **acoustic** ($\omega \to 0$ as $q \to 0$) and **optic** ($\omega \to$ finite as $q \to 0$);
- the dispersion relationships are periodic, such that $\omega(q) = \omega(q + j\frac{2\pi}{2a})$, where j is any integer.

The second point is a consequence of our studies of the reciprocal lattice in Chapter 2; we found that a periodic lattice in r-space implies that all particles and excitations must have dispersion relationships which are periodic in k-space. In our very simple example, the reciprocal lattice vectors are $G = 2\pi j/2a$, as the primitive lattice translation vector in r-space is $2a$. The nomenclature 'acoustic' comes from the fact that the lower branch tends towards the form $\omega = v_\phi q$, where v_ϕ is the speed of sound, as $q \to 0$; long wavelength acoustic phonons are nothing more than familiar sound waves.

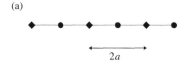

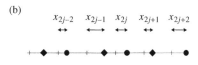

Fig. D.1 (a) A diatomic, linear (one-dimensional) chain, consisting of alternate masses M_1 (circles) and M_2 (diamonds) with equilibrium separation a (unit cell size $2a$). The atoms are shown in their equilibrium positions. (b) The atoms in the diatomic chain displaced from their equilibrium positions; x_{2j} is the displacement of the $2j$th atom.

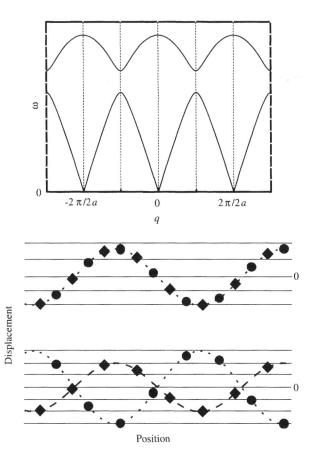

Fig. D.2 The phonon dispersion relationships predicted by eqn D.4.

Fig. D.3 The displacements of atoms M_1 (circles) and M_2 (diamonds) as a function of position for an acoustic mode (upper) and an optic mode (lower) of the same wavelength. The displacements are only meaningful at the atoms; the lines in between are a guide to the eye. Note that in this one-dimensional model the displacements are along the direction of the chain of atoms.

The term 'optic' is easier to fathom with the help of Fig. D.3 which shows the displacements of atoms M_1 (circles) and M_2 (diamonds) (obtained using eqns D.4 and D.2 and D.3) as a function of position for an acoustic mode (upper) and an optic mode (lower) of the same wavelength. (The displacements are only meaningful at the atoms; the lines in between are a guide to the eye.) Notice that the acoustic modes correspond to the displacement of the entire basis consisting of atoms M_1 and M_2, whereas in the optic mode, M_1 and M_2 vibrate *against* each other. If M_1 and M_2 are oppositely charged ions (as they will be in substances like GaAs, CdTe etc.), the optic phonon will therefore comprise a travelling polarisation wave which will interact strongly with electromagnetic radiation.

D.2.1 Extension to three dimensions

The model of phonons has thus far involved displacements along the chain direction. In three dimensions, the atoms can vibrate along the direction of propagation of the wave, and perpendicular to it. When a medium is isotropic, the three acoustic solutions for a given **q** (now a vector) can be chosen so that one branch (longitudinal mode) is polarised parallel to **q** and the other two are perpendicular to **q** (transverse modes). In anisotropic crystals, however, the polarisation vectors are not simply related to the direction of **q**

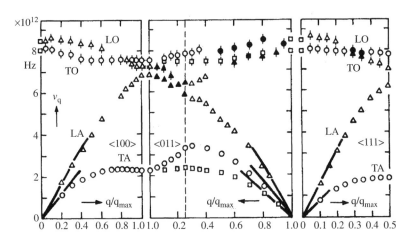

Fig. D.4 Phonon dispersion relationships of GaAs at a temperature of 296 K, deduced using neutron scattering (T = transverse, L = longitudinal, O = optic and A = acoustic). Data are plotted for **q** parallel to $<100>$, $<011>$ and $<111>$; the dashed line indicates the Brillouin zone boundary in the $<011>$ direction. In these high-symmetry directions, the two TO modes are degenerate, as are the two TA modes. Note that the TO and LO modes are *not* degenerate at $q = 0$, because GaAs is a polar material (the Ga ion has a positive charge and the As ion has a negative charge). As a result, the $q = 0$ LO mode has a higher associated electromagnetic energy density than the $q = 0$ TO mode (this is always true for polar crystals). However, in crystals with a nonpolar diatomic basis, such as Si and Ge, the TO and LO modes are degenerate at $q = 0$. The solid lines represent the velocities of sound (i.e. long-wavelength acoustic modes) deduced from the bulk elastic constants. (After G. Dolling and J.L.T. Waugh, *Lattice dynamics*, edited by R.F. Wallis (Pergamon, London, 1965).)

unless **q** is invariant under the symmetry operations of the crystal. Therefore the modes are not in general degenerate. In crystals of high symmetry (e.g. the various cubic types), there are many symmetry directions. Since the polarisation vectors of phonons are continuous functions of **q**, the branch that is longitudinal when **q** is along a symmetry direction tends to have a polarisation vector fairly close to **q**, even when **q** is not in a direction of symmetry. Similarly, the two transverse modes will be more-or-less perpendicular to **q** and in many cases degenerate. These considerations are illustrated by Fig. D.4.

Generalising to a basis consisting of l atoms (i.e. a primitive unit cell containing l atoms), for each value of **q** there will be $3l$ phonon modes. Each of the $3l$ branches will have a frequency ω which is a function of **q**, with the periodicity of the reciprocal lattice. Three of the branches will be acoustic ($\omega \rightarrow 0$ as $|\mathbf{q}| \rightarrow 0$), whilst the remaining $(3l - 3)$ will be optic (finite ω as $|\mathbf{q}| \rightarrow 0$), in exact analogy with the three translational and $(3l - 3)$ vibrational/rotational modes of a single, free basis 'molecule' consisting of l atoms. Therefore, in the case of a monatomic basis (as is the case in most elemental metals), only acoustic modes are present. This can also be seen by substituting $M_1 = M_2 = M$ in eqn D.4, which produces the simplified dispersion relationship

$$\omega = 2\left(\frac{\mathcal{F}}{M}\right)^{\frac{1}{2}} \left|\sin\left(\frac{qa}{2}\right)\right|. \tag{D.5}$$

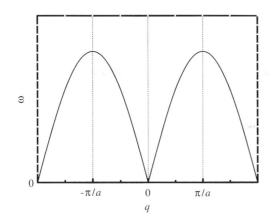

Fig. D.5 The phonon dispersion relationships predicted by eqn D.5.

Note that ω now has the k-space periodicity $2\pi/a$, as the fundamental periodicity in r-space is a, rather than $2a$. The dispersion relationship of eqn D.5 is shown in Fig. D.5

D.3 The Debye model

Debye derived a simple model of the heat capacity due to phonons. This involves a parameter called the Debye temperature, θ_D, which is nowadays extensively used as a convenient energy scale for phonons. In its simplest form, Debye's model assumes that the solid can be approximated by an isotropic, continuous medium. The vibrations of a continuous medium possess an isotropic, frequency-independent, linear dispersion relationship

$$\omega = v_\phi k, \quad (D.6)$$

where v_ϕ is the speed of sound. Taken with eqn B.9 ($g(k) \propto k^2$ in three dimensions), this implies that

$$g(\omega) \equiv \frac{dn}{d\omega} = \left(\frac{dn}{dk}\right)\left(\frac{dk}{d\omega}\right) = H\omega^2, \quad (D.7)$$

where H is a constant. The Debye model then constrains the total number of possible modes to be $3Nl$, where N is the number of primitive unit cells in the crystal and l is the number of atoms in the primitive cell, by imposing a maximum phonon frequency ω_D, i.e.

$$3Nl = \int_0^{\omega_D} H\omega^2 d\omega = \frac{H\omega_D^3}{3}. \quad (D.8)$$

We can therefore make the identification $H = 9Nl/\omega_D^3$, that is $g(\omega) = 9Nl\omega^2/\omega_D^3$. Hence, the total energy U of the phonon system is given by integrating the product of the energy of each mode $\hbar\omega$, $g(\omega)$ and a suitable distribution function, which in this case is the Bose–Einstein distribution function with $\mu = 0$. The result is

$$U = \int_0^{\omega_D} \hbar\omega \frac{9Nl}{\omega_D^3} \omega^2 \frac{1}{e^{\hbar\omega/k_B T} - 1} d\omega. \quad (D.9)$$

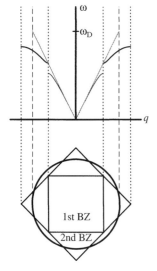

Fig. D.6 Comparison of Debye model and more realistic phonon dispersion curves. The upper part of the figure shows the acoustic and optic modes of a crystal with a simple diatomic basis, plotted as mode energy versus q (only one of each of the three possible acoustic and optic branches is shown, for clarity). In the Debye model, these are approximated by a straight line. The lower part of the figure shows the effective volumes of integration used in counting modes. In the Debye model, this is done by integrating over a sphere in k-space of diameter ω_D/v_ϕ, whereas the correct volume of integration, encompassing the natural symmetry of the crystal (see Chapter 2) and the k-space periodicity of the phonon modes is the Brillouin zone. In the diatomic crystal, the sphere must be of the same volume as the first two Brillouin zones in order to encompass the correct number of states. Note that a simple cubic lattice has been assumed, so that cross-section of the first Brillouin zone is square.

The heat capacity C due to phonons is given by differentiating the energy per unit volume $E/\mathcal{V}$, with $\mathcal{V}$ the volume of the crystal, with respect to T

$$C \equiv \frac{d\frac{E}{\mathcal{V}}}{dT} = \frac{9Nl}{\mathcal{V}\omega_D^3}\int_0^{\omega_D}\frac{\hbar^2\omega^4}{k_BT^2}\frac{e^{\hbar\omega/k_BT}}{(e^{\hbar\omega/k_BT}-1)^2}d\omega. \quad (D.10)$$

Making the substitutions $\theta_D = \hbar\omega_D/k_B$ and $x = \hbar\omega/k_BT$, we obtain

$$C = 9\frac{Nl}{\mathcal{V}}k_B\left(\frac{T}{\theta_D}\right)^3\int_0^{\theta_D/T}\frac{x^4e^x}{(e^x-1)^2}dx. \quad (D.11)$$

There are two limits that are of interest for us.

At low temperatures, (θ_D/T) becomes very large, so that the upper limit in eqn D.11 can be taken to be $x = \infty$. The integral is then just a dimensionless number, yielding

$$C = \frac{12}{5}\pi^4\frac{Nl}{\mathcal{V}}k_B\left(\frac{T}{\theta_D}\right)^3, \quad (D.12)$$

the **Debye T^3 law**. The phonon heat capacities of most materials obey this relationship rather well at low temperatures.

At sufficiently high temperatures, θ_D/T becomes small, so that all of the values of x also become small; hence, e^x can be approximated by $(1+x)$. Equation D.11 becomes

$$C = 9\frac{Nl}{\mathcal{V}}k_B\left(\frac{T}{\theta_D}\right)^3\int_0^{\theta_D/T}x^2dx = 3\frac{Nl}{\mathcal{V}}k_B. \quad (D.13)$$

This is the classical result for vibrating bound particles, being analogous to the result derived for free particles in Section C.1.4. It is in quite reasonable agreement with experimental observations. However, this is no big surprise; at high temperatures, all vibrational modes are excited, and so the heat capacity will be largely model-independent.

The Debye theory does not describe experimental data very well between these limits. The problem occurs because

- we have approximated the phonon dispersion relationships by $\omega = v_\phi k$, when in fact they are much more complex than this (see upper part of Fig. D.6);
- as a consequence, our simple expression $g(\omega) \propto \omega^2$ will be incorrect;
- we have in effect counted modes by integrating over a sphere in k-space of diameter ω_D/v_ϕ, whereas in fact the correct volume of integration, encompassing the natural symmetry of the crystal (see Chapter 2) and the k-space periodicity of the phonon modes is the Brillouin zone (see lower part of Fig. D.6).

The Debye model works well at temperatures $T \ll \theta_D$ because only low-energy phonons (i.e. with energies $\hbar\omega \sim k_BT$) will be present. At low energies, the phonon dispersion relationships are approximately linear; as modes higher in energy will not be excited, the exact form of the integration volume in k-space will not matter, as long as it encompasses the correct number of states. At a deeper level, low-energy phonons also have small k, and hence very long wavelengths, $\gg a$, the typical atomic spacing. A long wavelength vibrational mode will therefore not 'perceive' (if the anthropomorphism will be forgiven) the fine structure on a scale $\sim a$ of the crystal, and hence behave as though it moves in a continuous medium. This is of course exactly our starting assumption!

D.3.1 Phonon number

In the discussion of the scattering of electrons by phonons, it is useful to be able to find approximate expressions which describe the temperature dependence of the number of scattering events (which we expect to be determined by the number of phonons), and the average energy of the phonons involved. The number of phonon modes N_{ph} excited at a temperature T may be obtained by integrating the product of $g(\omega)$ and the distribution function;

$$N_{ph} = \int_0^{\omega_D} \frac{9Nl}{\omega_D^3} \omega^2 \frac{1}{e^{\hbar\omega/k_BT} - 1} d\omega. \tag{D.14}$$

Similarly, eqn D.9

$$U = \int_0^{\omega_D} \hbar\omega \frac{9Nl}{\omega_D^3} \omega^2 \frac{1}{e^{\hbar\omega/k_BT} - 1} d\omega.$$

gives the total energy of the phonon system. As before, we make the substitution $x = \hbar\omega/k_BT$ in both equations, to yield

$$N_{ph} = \frac{9Nl}{\omega_D^3} \left\{ \frac{k_BT}{\hbar} \right\}^3 \int_0^{\theta_D/T} \frac{x^2}{e^x - 1} dx \tag{D.15}$$

and

$$U = \frac{9Nl}{\omega_D^3} k_BT \left\{ \frac{k_BT}{\hbar} \right\}^3 \int_0^{\theta_D/T} \frac{x^3}{e^x - 1} dx. \tag{D.16}$$

It is useful to examine two limits.

Low temperatures. In the limit that $T \ll \theta_D$, the upper limit of the integrals can be set to infinity, yielding

$$N_{ph} = \frac{9Nl}{\omega_D^3} \left\{ \frac{k_BT}{\hbar} \right\}^3 \int_0^{\infty} \frac{x^2}{e^x - 1} dx \tag{D.17}$$

and

$$U = \frac{9Nl}{\omega_D^3} k_BT \left\{ \frac{k_BT}{\hbar} \right\}^3 \int_0^{\infty} \frac{x^3}{e^x - 1} dx \tag{D.18}$$

As in the derivation of the Debye heat capacity (see eqn D.11) the integrals in eqns D.17 and D.18 are just dimensionless numbers, yielding the important results for the number of phonons

$$N_{ph} \propto T^3 \tag{D.19}$$

and the average energy per phonon

$$\frac{U}{N_{ph}} = \xi k_BT \tag{D.20}$$

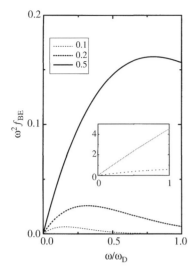

Fig. D.7 The distribution function $f_{BE} = \frac{1}{e^{\hbar\omega/k_B T}-1}$ multiplied by ω^2 plotted against ω. The main figure shows curves for temperatures $(T/\theta_D) = 0.1, 0.2, 0.5$, and the inset shows $(T/\theta_D) = 1, 5$. When T is a small fraction of θ_D, Debye phonons have a 'black-body-like' distribution, with a peak just above $\omega \sim k_B T/\hbar$, whereas at large temperatures, the distribution tends to a straight line, cut off at ω_D.

where ξ is a dimensionless number of order 1. The latter result occurs because phonons have a 'black-body' distribution somewhat analogous to that of photons, which peaks close to $k_B T$ at low temperatures; this is illustrated in Fig. D.7, which shows the Debye density of states multiplied by the distribution function, $g(\omega)\frac{1}{e^{\hbar\omega/k_B T}-1}$. The latter quantity gives a measure of the number of phonon states excited between ω and $\omega + d\omega$ at a temperature T.

High temperatures. For temperatures at which θ_D becomes small, e^x can be approximated by $(1+x)$. Equations D.15 and D.16 become

$$N_{ph} = \frac{9Nl}{\omega_D^3}\left\{\frac{k_B T}{\hbar}\right\}^3 \int_0^{\theta_D/T} x\,dx = \frac{9}{2}Nl\frac{T}{\theta_D} \qquad (D.21)$$

(i.e. $N_{ph} \propto T$) and

$$U = \frac{9Nl}{\omega_D^3}k_B T\left\{\frac{k_B T}{\hbar}\right\}^3 \int_0^{\theta_D/T} x^2\,dx = 3Nlk_B T, \qquad (D.22)$$

yielding an average phonon energy of $U/N_{ph} = 2k_B\theta_D/3$. At high temperatures, the Debye phonon system becomes very different from the photon black-body distribution; the latter has no high frequency cut-off, whereas the phonon frequencies have an upper limit of ω_D (see Fig. D.7).

D.3.2 Summary; the Debye temperature as a useful energy scale in solids

I hope that I have shown that the Debye model provides a useful approximation which enables the phonon contribution to the heat capacity to be derived at low temperatures, and which provides reasonable estimates for the phonon number and typical phonon energy. The relevant energy scale in the Debye model is given by the Debye temperature, θ_D. Figure D.6 also shows that the Debye phonon frequency ω_D provides a reasonable estimate of the most energetic phonon in a material; it is therefore a guide to the hardness of a material, because a harder material will have stiffer bonds between the component atoms or ions, and therefore more energetic phonons. For this reason, the Debye temperature θ_D is a very useful guide to the properties of a material which relate to its phonons. We shall find that the scattering of electrons by phonons can be described simply in the temperature regions $T \ll \theta_D$ and $T >\sim \theta_D$ (see Chapter 9); θ_D will also crop up as an energy scale in superconductivity (see Chapter H). Some typical Debye temperatures are shown in Table D.1.

D.3.3 A note on the effect of dimensionality

It is an interesting diversion to examine the Debye model in one dimension and two dimensions. It is left as an exercise for the reader to follow the working of Section D.3 with the following alterations;

$$2 \text{ dimensions}: \quad g(\omega) \propto \omega$$

$$1 \text{ dimension}: \quad g(\omega) = \text{constant}$$

Table D.1 Debye temperatures for selected substances. Note the trend for θ_D to decrease as one descends the periodic table. In the cases where two values are given, there are some discrepancies between the published values. The carbon data refer to diamond. (Extracted from J. de Launay, *Solid state physics*, **vol. 2**, edited by F. Seitz and D. Turnbull (Academic Press, New York, 1956), J.T. Lewis et al., *Phys. Rev.* **161**, 877 (1967) and Landholt-Börnstein, *Numerical data and functional relationships in science and technology*, New Series, Group III, Volume 17, edited by O. Madelung (Springer, Berlin, 1982).)

Substance	$\theta_D(K)$	Substance	θ_D (K)
Li	400	Na	150
K	100	Be	1000
Mg	318	Ca	230
B	1250	Al	394
Ga	240	In	129
C	1860	Si	625
Ge	360	Pb	88
As	285	Sb	200
Bi	120	A	85
Ne	63	Cu	315
Ag	215	Au	170
Zn	234	Cd	120
Fe	420	Co	385
Pd	275	Pt	230
La	132	Gd	152
NaF	492	NaCl	321
NaBr	224	KF	336
KCl	231	KBr	173
InSb	203, 208	InP	321
InAs	247, 262	GaSb	266
GaP	445, 468	CdTe	158
HgTe	147, 141.5		

(see eqn B.10; remember that the Debye phonons have $\omega = v_\phi k$). The total possible number of modes is $2Nl$ in two dimensions, and Nl in one dimension. You should find that the phonon heat capacity C for a one-dimensional system is proportional to T at low temperatures, and becomes Nlk_B/d at high temperatures (d = specimen length). Similarly, for a two-dimensional system, $C \propto T^2$ at low temperatures, and $C = 2Nlk_B/A$ at high temperatures (A = specimen area). In fact, it is possible to show that, at low temperatures, $C \propto T^j$ for a j-dimensional system. This result has been checked experimentally for the case of $j = 1$; see A. Potts *et al.*, in *High magnetic fields in semiconductor physics III: quantum Hall effect, transport and optics*, ed. G. Landwehr, Springer Series in Solid-State Sciences, **101**, 325 (1992).

Further reading

For an alternative discussion of the phonon heat capacity, see e.g. *Introduction to solid state physics*, by Charles Kittel, seventh edition (Wiley, New York, 1996), Chapter 5, *Three phases of matter*, by Alan J. Walton, second edition (Clarendon Press, Oxford, 1983), Chapter 10, *Solid state physics*, by G. Burns (Academic Press, Boston, 1995), Section 11.5, *Solid state physics*, by N.W Ashcroft and N.D. Mermin (Holt, Rinehart and Winston, New York, 1976), Chapter 23. Chapter 2 of *Quantum theory of the solid state*, by Joseph Callaway (Academic Press, San Diego, 1991) is recommended as a very good introduction to more sophisticated treatments of phonons.

The Bohr model of hydrogen

E

E.1 Introduction

Impurities and excitons in semiconductors behave in a very similar way to the hydrogen atom. In the hydrogen atom, a single electron orbits a nucleus consisting of a single proton. The application of the Schrödinger equation to the hydrogen atom is explored in numerous quantum mechanics and 'modern' physics books, a couple of which are mentioned in Further Reading. However, the characteristic energy and length scales of the hydrogen atom can be obtained easily and with very good accuracy using the Bohr model. The Bohr model also has the virtue that modifications due to the motion of both nucleus and electron and due to the electromagnetic environment (the relative permittivity) can be made in an easily-understood manner.

First of all we consider an electron of mass m_e and charge $-e$ moving at a speed v around a fixed (immovable) nucleus of charge $+e$ in a circular orbit of radius r;[1] both are *in vacuo*. In order for such an orbit to occur, the Coulomb force between the nucleus and electron must provide the centripetal acceleration characteristic of a circular path;

$$\frac{e^2}{4\pi\epsilon_0 r^2} = \frac{m_e v^2}{r} \tag{E.1}$$

Bohr introduced quantum mechanics by proposing that the electron's angular momentum must be quantised in units of $\hbar$, i.e.

$$m_e v r = \frac{nh}{2\pi} = n\hbar, \tag{E.2}$$

where n is an integer. Combining this with eqn E.1, we obtain the radii r_n of the allowed orbits

$$r_n = \frac{\epsilon_0 h^2 n^2}{\pi m_e e^2} = a_0 n^2 = 0.529 n^2 \text{ Å}, \tag{E.3}$$

where a_0 is known as the *Bohr radius*. The total energy E_n of the electron is obtained by summing its kinetic energy $\frac{1}{2}m_e v^2$ and potential energy $-e^2/(4\pi\epsilon_0 r)$ to give

$$E_n = -\frac{m_e e^4}{8\epsilon_0^2 h^2 n^2} = -\frac{13.6}{n^2} \text{ eV}. \tag{E.4}$$

E.1	Introduction	191
E.2	Hydrogenic impurities	192
E.3	Excitons	192

[1] Classically, one would expect an orbiting (i.e. accelerating) electron to continuously emit radiation and therefore to lose energy. For this to happen, the electron would have to have a continuous distribution of lower energy states into which it could fall. Bohr realised that the energy states were quantised and discrete, and therefore that radiation could only come out in quantised 'lumps', the characteristic photon energies which give rise to the lines of the atomic spectra.

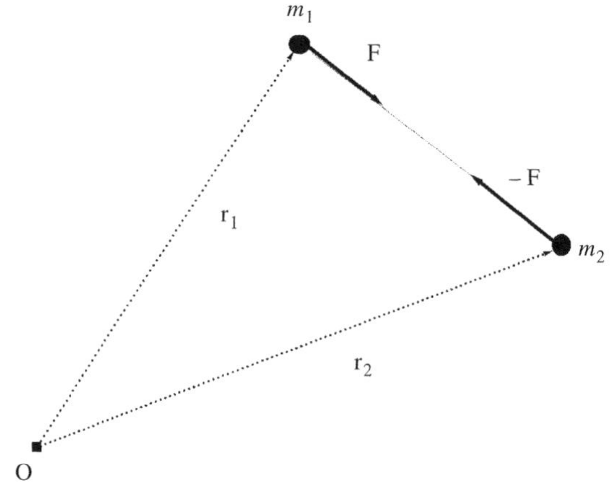

Fig. E.1 Two moving particles, masses m_1 and m_2 and positions $\mathbf{r}_1$ and $\mathbf{r}_2$ interacting via a force $\mathbf{F}$ which only depends on their relative separation.

E.2 Hydrogenic impurities

In considering hydrogenic impurities (see Section 6.4.4), a carrier (electron or hole) of charge $\mp e$ orbits a fixed parent ion (donor or acceptor) of charge $\pm e$. Our assumption about the fixed 'nucleus' still holds, and the magnitudes of the charges involved are the same as in the hydrogen model. However, in the case of the impurity, the carrier moves within a medium of relative permittivity ϵ_r and in a band with a characteristic effective mass m^*. The substitutions $\epsilon_0 \to \epsilon_0 \epsilon_r$ and $m_e \to m^*$ must be made in eqn E.4 to yield

$$E(n) = \frac{m^*}{m_e} \frac{1}{\epsilon_r^2} \times \frac{13.6 \text{ eV}}{n^2}. \tag{E.5}$$

Similar substitutions can be made in eqn E.3 to yield an effective Bohr radius, a^*.

E.3 Excitons

In an exciton, an electron and a hole orbit each other; both have effective masses of a similar order of magnitude, and neither is fixed.[2] The general situation is shown in Fig. E.1, which depicts two moving particles, 1 and 2, of masses m_1 and m_2 and positions $\mathbf{r}_1$ and $\mathbf{r}_2$. These interact via a force $\mathbf{F}$, the magnitude of which only depends on their relative separation. In other words

$$\mathbf{F} = \mathbf{F}(|\mathbf{r}_1 - \mathbf{r}_2|).$$

As every action has an equal and opposite reaction, the force exerted by 1 on 2 is equal and opposite to that exerted by 2 on 1.

The equations of motion are

$$m_1 \frac{d^2 \mathbf{r}_1}{dt^2} = \mathbf{F}(|\mathbf{r}_1 - \mathbf{r}_2|) \tag{E.6}$$

and

$$m_2 \frac{d^2 \mathbf{r}_2}{dt^2} = -\mathbf{F}(|\mathbf{r}_1 - \mathbf{r}_2|). \tag{E.7}$$

---

[2] This is of course also true for the hydrogen atom. However, in the case of hydrogen, the effect is not very noticeable, as the mass of the proton is just over 1800 times m_e; it results in tiny but observable corrections to the energies E_n obtained.

At first this looks like a nasty situation, as the equations are coupled. However, we can divide eqn E.6 by m_1 and eqn E.7 by m_2 and subtract, to give

$$\frac{d^2\mathbf{r}_1}{dt^2} - \frac{d^2\mathbf{r}_2}{dt^2} = \left(\frac{1}{m_1} + \frac{1}{m_2}\right)\mathbf{F}(|\mathbf{r}_1 - \mathbf{r}_2|). \quad (E.8)$$

Remembering that

$$\frac{d^2\mathbf{r}_1}{dt^2} - \frac{d^2\mathbf{r}_2}{dt^2} \equiv \frac{d^2(\mathbf{r}_1 - \mathbf{r}_2)}{dt^2}$$

and making the substitutions

$$\frac{1}{\mathcal{M}} = \left(\frac{1}{m_1} + \frac{1}{m_2}\right) \quad (E.9)$$

and

$$\mathbf{R} = \mathbf{r}_1 - \mathbf{r}_2$$

eqn E.8 becomes

$$\mathcal{M}\frac{d^2\mathbf{R}}{dt^2} = \mathbf{F}(\mathbf{R}). \quad (E.10)$$

We have succeeded in making a single differential equation in terms of the variable $\mathbf{R}$, the relative separation of the particles 1 and 2. The equation looks *exactly* like that of a single particle of mass $\mathcal{M}$, position $\mathbf{R}$ moving under the action of a force $\mathbf{F}$ which depends only on $\mathbf{R}$. The quantity $\mathcal{M}$ defined by eqn E.9 is known as the *reduced mass*. Therefore, to take account of the fact that the nucleus (mass M) of hydrogen is *not* immobile, but is free to move, all we need to do is to make the substitution $m_e \to \mathcal{M}$, where $(1/\mathcal{M} = (1/M + 1/m_e)$; the rest of the algebra will work exactly as before.[3]

In the case of the exciton (a bound electron–hole pair; see Section 6.3.4), $\mathcal{M}$ is replaced by an effective reduced mass μ^* given by

$$\frac{1}{\mu^*} = \frac{1}{m_c^*} + \frac{1}{m_{hh}^*} \quad (E.11)$$

where m_c^* is the effective mass of the electron in the conduction band and m_{hh}^* is the mass of the heavy hole in the valence band. Therefore the substitutions $\epsilon_0 \to \epsilon_0\epsilon_r$ and $m_e \to \mu^*$ must be made in eqn E.4 to yield the energies of the excitonic states;

$$E(n) = \frac{\mu^*}{m_e}\frac{1}{\epsilon_r^2} \times \frac{13.6 \text{ eV}}{n^2}. \quad (E.12)$$

Similar substitutions can be made in eqn E.3 to yield an effective Bohr radius, a^*, of the exciton.

[3] As I have already remarked, in the case of hydrogen, $M \gg m_e$, so that $\mathcal{M} \approx m_e$. This results in in a tiny shift in the predicted energies E_n.

Further reading

The Bohr model of the hydrogen atom is treated by most elementary quantum-mechanics texts; see *e.g. Quantum physics of atoms, molecules, solids, nuclei and particles*, by R. Eisberg and R. Resnick (Wiley, New York, 1985), Sections 4.5 to 4.12. Most quantum mechanics texts give a more rigorous solution of the Schrödinger equation for the hydrogen atom; see Chapter 7 of Eisberg and Resnick or Chapter 12 of *Quantum mechanics*, by Stephen Gasiorowicz (Wiley, New York, 1974).

Experimental considerations in measuring resistivity and Hall effect

F.1	Introduction	194
F.2	The four-wire method	194
F.3	Sample geometries	196
F.4	The van der Pauw method.	197
F.5	Mobility spectrum analysis	198
F.6	The resistivity of layered samples	198

[1] See, for instance, the resistance data for metals and semiconductors in Chapter 9. As another example, a routine characterisation measurement on a semiconductor sample (used to asses the quality of a batch of material) will involve at the very least measurements at room temperature ($\sim$ 300 K), the boiling point of liquid nitrogen at atmospheric pressure (77 K) and the boiling point of liquid helium (4.2 K) at atmospheric pressure.

[2] This is not just a consideration of space within the cryostat; in an accurate measurement of resistivity, it is best to work with high-quality single crystals, which usually have dimensions $\sim$ 1 mm (bigger crystals tend to contain inhomogeneities, grain boundaries etc.). Furthermore, if the resistivity measurement is used to characterise a batch of material intended for something much more lucrative (e.g. making devices), it is obviously desirable to only require a small fraction of the batch for the characterisation!

F.1 Introduction

Resistivity and Hall effect measurements are presented in several places in this book, and so it is worthwhile to summarise the techniques used to acquire such data.

In the following sections, it is worth remembering that a useful resistivity measurement will probably involve subjecting the sample to a wide range of temperatures.[1] Furthermore, a measurement of the Hall effect requires that the sample (and the means of controlling its temperature- usually a cryostat or oven) be placed within a magnet of some kind. Thus, the sample is usually at some distance from the measuring equipment, and must be connected to it using long wires. These wires must not perturb the temperature of the sample by, for example, acting to conduct heat to or from it; this implies that the wires must be small and of a low thermal conductivity. Remembering that a low thermal conductivity implies a low electrical conductivity (see Chapter 9), the wires must therefore have a rather large electrical resistance.

Virtually all methods measure resistivity by driving a current between one pair of contacts on the sample and measuring the resultant voltage across another pair of contacts. This is the so-called **four-wire method**. The samples are almost always small[2] so that one must employ small currents to avoid the measurement heating the sample up (and therefore changing the sample's resistance). This means that the voltages to be measured are small, and that electrical interference and noise and other sources of voltage are liable to disturb the measurement.

F.2 The four-wire method

In the four-wire method, separate pairs of wires are used for providing the current through the sample and measuring the voltage across it. A simplified circuit diagram is shown in Fig. F.1 (upper diagram). The inferior alternative to the four-wire method is to use the same leads for providing the current and measuring the voltage (the **two-wire method**, shown in the lower part of Fig. F.1). However, this is very undesirable; as we have seen, the wires are liable to be of a considerable (and poorly-characterised) resistance, which we shall call $r_{\rm wire}$. Furthermore, the electrical contacts on semiconductor samples often possess large resistances $r_{\rm con}$. Therefore, the two-wire method will measure a voltage

$$V_{\text{meas}} = Ir_{\text{con}} + Ir_{\text{wire}} + IR;$$

the contribution of the sample, of resistance R, is likely to be masked or disturbed by the other components.

The four-wire method gets around this problem because typical voltage-measuring devices have very high input impedances. If we assume, for a moment, that the impedance of the voltage measuring device in Fig. F.1 is infinite, then *no* current will flow in the voltage-measuring wires and contacts; therefore no voltage will be dropped across them, and the voltage measured will be the true voltage dropped across the sample. In reality, most non-specialised voltage-measuring devices (digital multimeters, lock-in amplifiers) have an effective input impedances of ~ 100 MΩ. Nevertheless, as long as the resistance of the sample does not go much above ~ 1 MΩ, very little current will flow through the voltage-measuring part of the circuit, and the voltage measured will be mainly due to the sample.

Further problems in the measurement of resistance can result from the presence of **thermal voltages**, caused by the junctions between the different metals in the circuitry being at different temperatures (the mechanism is the same as that of the **thermocouple**).[3] Let us suppose that we are trying to measure a resistance R using a current I. The measured voltage V_{meas1} that we would get would be

$$V_{\text{meas1}} = IR + V_{\text{therm}}, \quad (\text{F.1})$$

where V_{therm} is the contribution from thermal voltages. One can get around this by reversing the current; in this case the measured voltage V_{meas2} would be

$$V_{\text{meas2}} = -IR + V_{\text{therm}}, \quad (\text{F.2})$$

as the thermal voltages do not depend on the current direction. The true resistance could be deduced from

$$R = \frac{V_{\text{meas1}} - V_{\text{meas2}}}{2I}. \quad (\text{F.3})$$

In many cases, experimental physicists use low-frequency alternating currents to get around this problem, with a lock-in amplifier to measure the voltages. However, this is not a panacea, as the electrical contacts on samples of materials such as semiconductors often have associated capacitances.[4] Moreover, the wiring of a poorly-designed apparatus may contain stray inductances and capacitances. Very careful checks of the relative phase of the current and sample voltage must be made to ensure that these contributions are not masking the true behaviour of the sample. Papers in the literature often contain disclaimers along the lines of 'great care was taken to ensure that the measured resistance was neither frequency nor current dependent and that no heating of the sample occurred'.

A rule of thumb for assessing whether the current is heating the sample or not is to double the current and check whether the measured resistance is the same; if it is not, then the current is probably warming the sample up.[5] Reduce the current and try again until consistent values are obtained.

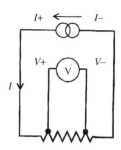

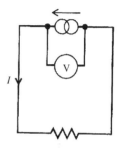

Fig. F.1 Above: schematic of the *four–wire* method of measuring resistance. The lower figure shows the inferior *two-wire* method; the measured resistance includes contributions from the resistances of the wires and the contacts on the sample. (In both cases, the resistor symbol represents the sample.)

[3] See, for example, page 156 of *Experimental low temperature physics*, by A.J. Kent (Macmillan, London 1993).

[4] See, for example, *Physics of semiconductor devices*, by S.M. Sze (Wiley-Interscience, New York, 1981).

[5] There are more exotic mechanisms for generating a nonlinear current–voltage relationship. However, it is worth eliminating the obvious before staking your reputation and claiming a 'new effect'.

196 Experimental considerations in measuring resistivity and Hall effect

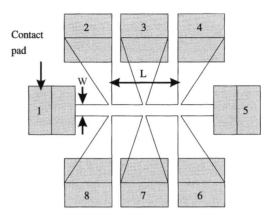

Fig. F.2 Typical geometry of a sample for resistivity measurements in the case that the material being studied is a thin film grown on an insulating substrate. The electrically-active top layer is etched into the shape outlined in black, leaving the substrate as a support. The central bar (typical widths will be $W \sim 10 - 100~\mu\text{m}$; $L \sim 100 - 500~\mu\text{m}$) is the region under study; the rectangles at each end are the current contacts and the triangles on each side form the voltage-measuring contacts. The grey areas represent evaporated metal films which cover the edges of the contacts and the some of the remaining substrate. It is customary to anneal such contacts so that some of the metal diffuses into the underlying sample, thus ensuring a good electrical connection. Wires can then be bonded onto the metal pads. Note that the contact regions are kept well away from the central region under study. (Figure provided by Phil Gee, Oxford Instruments plc.)

F.3 Sample geometries

In studying bulk materials (as opposed to layered materials), measurements are made on rod-shaped samples (single crystals, needles, rods sawn from single crystals etc.). Contacts are soldered, evaporated or glued (using electrically-conductive paint) to the faces of the rod in arrangements that approximate that shown in Fig. 10.1, so that simultaneous measurements of V_x ($\propto$ the resistivity component ρ_{xx}) and V_y (the Hall voltage, $\propto \rho_{xy}$) are feasible. As the samples are small and the contacts are often applied by hand, these contact arrangements are often less than ideal. A particular problem stems from the fact that unless the contacts used to measure V_y are exactly opposite each other, then some component proportional to ρ_{xx} will interfere with the measurement. This can be overcome by reversing the magnetic field; ρ_{xy} will change sign, whereas ρ_{xx} will not (see Chapter 10), so that the difference of the two measured voltages will give a much more accurate value for the true Hall voltage.[6]

[6]See H.H. Wieder, *Laboratory notes on electrical and galvanomagnetic measurements* (Elsevier, Amsterdam, 1979) for another method of overcoming this problem using an alternating magnetic field combined with an alternating current.

In the case of semiconductor heterostructures, or any thin layer grown on an insulating substrate, the techniques of photolithography and/or chemical or mechanical etching can be used to produce a Hall-bar sample of almost ideal geometry; the top layer is etched into a shape such as that in Fig. F.2, but the supporting substrate is left unetched to ensure mechanical strength.

Note that the metal pads to which one can connect wires run over the edge of the layer onto the sample. This is to avoid a particular problem when semiconductor heterostructures containing two-dimensional electron or hole systems are measured in high magnetic fields. As is shown in Chapter 11, under certain conditions and at low temperatures, both σ_{xx} and ρ_{xx} can become

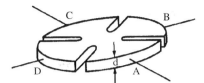

Fig. F.3 The 'clover-leaf' geometry of a van der Pauw sample of thickness d.

zero over extended regions of magnetic field. If a contact is placed so that it is completely surrounded by the two-dimensional carrier system, then it will become electrically isolated when $\sigma_{xx} \to 0$, leading to spurious results. In contrast, the measurement is generally found to be well-behaved if the metal pad is not completely surrounded by the two-dimensional carrier system, i.e. if it runs over the edges of the electrically-active region onto the substrate, as shown in Fig. F.2.[7]

[7] A discussion of some of the mechanisms leading to this behaviour is given in G. Nachtwei, *Physica E* **4** 79 (1999) and references therein.

F.4 The van der Pauw method.

An elegant method was derived by van der Pauw[8] for measuring the resistivity components of samples with four contacts. Ideally, the sample should be of the clover-leaf shape shown in Fig. F.3, but the current author and others have applied the technique to quite arbitrarily-shaped samples with some success. The method for obtaining the zero-magnetic-field resistivity is as follows (refer to Fig. F.3 for the contact labels).

[8] L.J. van der Pauw, *Philips Research Reports* **16**, 187 (1961).

(1) A current I_{AB} is driven from contact A to contact B; the resulting voltage V_{CD} is measured between contacts C and D. The resistance $R_{AB,CD} = |V_{CD}|/I_{AB}$ is computed.

(2) A current I_{BC} is driven from contact B to contact C; the resulting voltage V_{AD} is measured between contacts A and D. The resistance $R_{BC,AD} = |V_{AD}|/I_{BC}$ is computed.

(3) The resistivity ρ is calculated using

$$\rho = \frac{\pi}{\ln 2} d \frac{R_{AB,CD} + R_{BC,AD}}{2} f,$$

where the function f dependends on the ratio $R_{AB,CD}/R_{BC,AD}$ (it is shown in Fig. F.4) and d is the sample thickness. The received wisdom is that one tries making new contacts if $R_{AB,CD}/R_{BC,AD}$ is bigger than ~ 3.

The Hall effect is often measured using such a sample by following the recipe below.

(1) A current I_{AC} is driven from A to C and the voltage between B and D ($V_{BD}(B)$) is measured in finite field B and at $B = 0$ to give $R_{H1} = (V_{BD}(B) - V_{BD}(B=0))d/(I_{AC}B)$.

(2) A current I_{BD} is driven from B to D and the voltage between A and C ($V_{AC}(B)$) is measured in finite field B and at $B = 0$ to give $R_{H2} = (V_{AC}(B) - V_{AC}(B=0))d/(I_{BD}B)$.

(3) The Hall coefficient is the average of these values; $R_H = \frac{1}{2}(R_{H1} + R_{H2})$.

Many researchers employ a 'belt and braces' approach, repeating steps 1 and 2 at negative magnetic field ($-B$) and averaging all four values to get R_H.

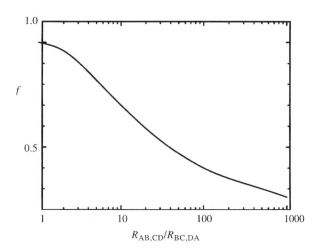

Fig. F.4 The function f used in the van der Pauw technique, plotted as a function of $R_{AB,CD}/R_{BC,AD}$.

F.5 Mobility spectrum analysis

The Hall effect and magnetoresistance are often very difficult to analyse if several species of carrier are present simultaneously in a sample (for example a heterostructure containing light and heavy holes might also possess a surface accumulation layer containing electrons). Beck and Anderson developed an elegant method called *mobility spectrum analysis* which solves this problem.[9] The method applies numerical transformations to a series of simultaneous measurements of ρ_{xx} and ρ_{xy}, yielding a plot of carrier density versus mobility referred to as a *mobility spectrum*; each species of carrier leads to a peak on this graph. The method has been applied successfully to a number of different systems.[10]

F.6 The resistivity of layered samples

Layered and quasi-two-dimensional materials are at present a 'hot' research topic; examples include the high-T_c cuprate superconductors, the organic molecular metals and the layered magnetic oxides (see Chapter 7). Most resistivity studies of these materials involve measurements of ρ_{zz}, the resistivity component in the direction perpendicular to the highly conductive planes.[11] The reasons for this are twofold. First, the resistivity in the interplane direction is usually several orders of magnitude larger (and therefore easier to measure) than that in the intraplane direction, reflecting the extreme anisotropy of the bandstructure. Second, the single crystals tend to be irregular in shape, so that the apparent measured in-plane resistivity, ρ_m, is a combination of all of the components of the resistivity tensor. For example, in the simplest measurement geometry, i.e. four contacts on the same large face (parallel to the high conductivity planes) of the crystal, the current is concentrated near this face, leading to an apparent increase of the in-plane resistivity which has misled many research groups in the past. Buravov *et al.*[12] made a careful study of such effects, noting that for samples in which $s/l \geq (\rho_{\parallel}/\rho_{zz})^{\frac{1}{2}}$, where s and l are the sample thickness and distance between the current contacts respectively,

[9] See W.A. Beck and J.R. Anderson, *J. Appl. Phys.* **62**, 541 (1987).

[10] See e.g. S.P. Svensson *et al.*, *J. Crystal Growth*, **111**, 450 (1991) and W.A. Beck *et al.*, *J. Vac. Sci. Technol. A*, **6**, 2772 (1988).

[11] See e.g. John Singleton, *Reports on Progess in Physics* **63**, 1111 (2000).

[12] L.I. Buravov, *et al.*, *J. Phys. I France* **4**, 441 (1994).

and $\rho_{||} \approx \frac{1}{2}(\rho_{xx} + \rho_{yy})$, the measured resistivity is $\rho_m \approx 2(s/l)(\rho_{||}\rho_{zz})^{\frac{1}{2}}$. By contrast, if one places one current contact and one voltage contact on each large face of a crystal, the measured four-wire resistance will be almost exactly proportional to ρ_{zz}, irrespective of geometrical imperfections of the contact positions.

Further reading

One of the reasons for writing this Appendix is that there is a dearth of useful literature concerning resistivity measurements. Those interested in carrying out such measurements are referred to the papers cited above. Some useful hints and references are also given in Chapter 4^{13} of *Semiconductor physics*, by K. Seeger (Springer, Berlin, 1991) and in *The Hall effect and its applications*, edited by C.L. Chien and C.R. Westgate (Plenum, New York, 1980). Cryogenic arrangements used for (magneto)resistance experiments are described in *Experimental low temperature physics*, by A.J. Kent (Macmillan, London 1993) and *Matter at low temperatures*, by P.V.E McLintock, D.J. Meredith and J.K. Wigmore, (Blackie, Glasgow, 1984).

[13] This chapter also includes a description of the Corbino geometry, a two-contact method for measuring the magnetoresistance of materials which has had some important applications.

G Canonical momentum

In classical mechanics the force **F** on a particle with charge q moving with velocity **v** in an electric field **E** and magnetic field **B** is

$$\mathbf{F} = q(\mathbf{E} + \mathbf{v} \times \mathbf{B}). \tag{G.1}$$

With this familiar equation, one can show how the momentum of a charged particle in a magnetic field is modified. Using $\mathbf{F} = m\dot{\mathbf{v}}$, $\mathbf{B} = \nabla \times \mathbf{A}$ and $\mathbf{E} = -\nabla V - \partial \mathbf{A}/\partial t$, where V is the electric potential, $\mathbf{A}$ is the magnetic vector potential and m is the mass of the particle, eqn G.1 can be rewritten

$$m\frac{d\mathbf{v}}{dt} = -q\nabla V - q\frac{\partial \mathbf{A}}{\partial t} + q\mathbf{v} \times (\nabla \times \mathbf{A}). \tag{G.2}$$

As there is a vector identity which leads to

$$\mathbf{v} \times (\nabla \times \mathbf{A}) = \nabla(\mathbf{v} \cdot \mathbf{A}) - (\mathbf{v} \cdot \nabla)\mathbf{A} \tag{G.3}$$

we have

$$\frac{d}{dt}(m\mathbf{v} + q\mathbf{A}) = -q\nabla(V - \mathbf{v} \cdot \mathbf{A}). \tag{G.4}$$

Here, the $d\mathbf{A}/dt$ is the *convective derivative* of **A**, written as

$$\frac{d\mathbf{A}}{dt} = \frac{\partial \mathbf{A}}{\partial t} + (\mathbf{v} \cdot \nabla)\mathbf{A}, \tag{G.5}$$

which measures the rate of change of **A** at the location of the moving particle. Equation G.4 takes the form of Newton's second law (i.e. it reads 'rate of change of a quantity that looks like momentum is equal to the gradient of a quantity that looks like potential energy' and therefore motivates the definition of the **canonical momentum**

$$\mathbf{p} = m\mathbf{v} + q\mathbf{A} \tag{G.6}$$

and an effective potential energy experienced by the charged particle $q(V - \mathbf{v} \cdot \mathbf{A})$ which is now velocity-dependent. The canonical momentum reverts to the familiar momentum $m\mathbf{v}$ in the case of no magnetic field, $\mathbf{A} = 0$. The kinetic energy remains equal to $\frac{1}{2}mv^2$ and this can therefore be written in terms of the canonical momentum as $(\mathbf{p} - q\mathbf{A})^2/2m$. This result will be used in many places in the book where the quantum mechanical operator associated with kinetic energy in a magnetic field is written $(-i\hbar\nabla - q\mathbf{A})^2/2m$.

Superconductivity

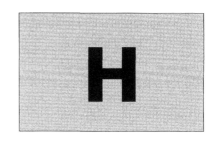

H.1 Introduction

Superconductors are materials whose electrical resistance drops to zero (or an immeasurably low value) on cooling below a well-defined critical temperature, T_c. The phenomenon of superconductivity is mentioned at several points in this book; it is therefore useful to have a simple explanation of the effect. I shall describe the model of superconductivity that applies to simple, elemental metals, which involves interactions between electrons mediated by phonons (other mechanisms will be alluded to under *Further reading* below). The evidence for the involvement of phonons was based primarily on two observations;

(1) elements which have a high T_c are generally of relatively low electrical conductivity at room temperature; as the electrical conductivity at room temperature is limited by phonons, this suggests that the electron-phonon interactions are relatively strong;

(2) if two isotopes of the same element are compared, T_c is found to be lower the heavier the isotope; this is known as the **isotope effect**.

H.1	Introduction	201
H.2	Pairing	201
H.3	Pairing and the Meissner effect	203

H.2 Pairing

Metals conduct electricity because they contain a large number of electrons which are relatively free to move through the regular array of positive ions that make up the crystal structure. However, as the electrons pass each ion, there will be a momentary attraction between them which will modify the way in which the ion vibrates (see Fig. H.1). As we have seen in Section D, the ions within a crystal do not vibrate in isolation; instead, they are connected by bonds to all of the ions around them; any change in the vibrational modes of the system is not local, but is passed on to other nearby ions. The vibration caused by the first electron will propagate through the crystal and can eventually interact with a second electron (Fig. H.1). In the language of field theory, this interaction is due to the exchange of a virtual phonon between the electrons.

It turns out that this interaction is weakly attractive; it is strong for electrons with equal and opposite wavevectors, and results in a momentary pairing of the electrons. A finite energy (usually referred to as the energy gap 2Δ, i.e. Δ per electron) is then required to break up the pair.[1] Why does this result in zero resistance? There are two ways of looking at this.

[1] Excellent evidence for the presence of the energy gap has been provided in a variety of experiments, including specific heat measurements, far-infrared/millimetre-wave reflectivity and transmission, ultrasound attenuation, and tunnelling experiments.

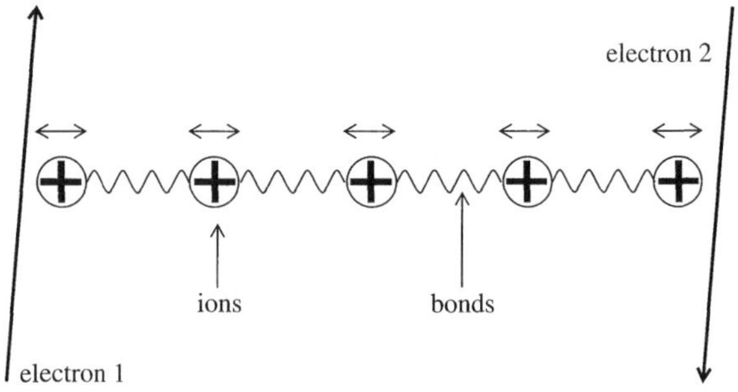

Fig. H.1 The pairing interaction between electrons occurs because the motion of electron 1 modifies the ionic vibrations and this in turn interacts with electron 2.

(1) Electrical resistivity occurs because the electrons scatter from impurities and lattice vibrations in the crystal, which have a characteristic size of order the ionic spacing $a \sim 0.2 - 0.3$ nm. Strong scattering only occurs when the de Broglie wavelength of the particle involved is less than or of the order of the size of the scatterer. The pair of electrons has no net wavevector (as the component electrons have equal and opposite wavevectors); in effect the pair has an infinite de Broglie wavelength. As a result scattering (and therefore resistivity) will be completely suppressed.

(2) A pair of electrons can only be scattered if the energy involved is sufficient to break it up into two single electrons. In general, at low temperatures this energy will not be available, and so the pair passes on, undeviated by scatterers.

The second point indicates the ways in which superconductivity can be suppressed so that the metal returns to its normal state; if energy sufficient to break up pairs is put in, then the superconductivity will cease.

- The energy could be thermal, i.e. heat the sample up until $k_\mathrm{B} T \sim \Delta$. We have already seen that the temperature below which superconductivity occurs is called T_c; we might consequently expect that $k_\mathrm{B} T_\mathrm{c} \sim \Delta$.
- Magnetic fields have an associated magnetic energy density and will suppress superconductivity; this occurs at a field B_c, known as the **critical field**.
- A sufficiently large current density (the **critical current density** J_c) will cause the superconductivity to break down (see if you can come up with a hand-waving explanation for this).

The development of the pairing hypothesis culminated in the Bardeen–Cooper–Schrieffer (BCS) theory, and this is the basis of our current understanding of superconductivity. The BCS theory predicts that $2\Delta \approx \frac{7}{2} k_\mathrm{B} T_\mathrm{c}$, quite close to the simple relationship derived above.

There are a number of analytically-tractable limits of BCS theory. For example, the *weak-coupling limit* gives the following expression for T_c:

$$T_\mathrm{c} = 1.13 \theta_\mathrm{D} e^{-1/(\lambda - \mu_\mathrm{C})}.$$

Here, λ is the electron–phonon coupling constant and μ_C is known as the Coulomb pseudopotential; it gives a measure of the Coulomb repulsion between the electrons which will oppose pairing. Note that the characteristic energy scale is θ_D, the Debye temperature (see Section D.3.2), again illustrating the involvement of phonons. The isotope effect can now be understood; eqn D.5 shows that the characteristic energy of phonons scales as $(\frac{\mathcal{F}}{M})^{\frac{1}{2}}$, where $\mathcal{F}$ is a 'spring constant' determined by the chemical bonds (and therefore the same for isotopes of the same element) and M is the atomic mass (and therefore different for different isotopes of the same element). Thus, a heavier isotope leads to a lower characteristic phonon energy, and hence a lower T_c.[2]

[2] Indeed, in a number of elements, $T_c \propto M^{-1/2}$.

H.3 Pairing and the Meissner effect

The fact that the pairs have zero momentum has another interesting consequence. As shown in Appendix G, the canonical momentum $\mathbf{p}$ of a particle of mass m, velocity $\mathbf{v}$ and charge q is given by

$$\mathbf{p} = m\mathbf{v} + q\mathbf{A}, \tag{H.1}$$

where $\mathbf{A}$ is the magnetic vector potential. We have already seen that $\mathbf{p} = 0$ for the superconducting pairs when $\mathbf{A} = 0$; we make the further assumption that $\mathbf{p} = 0$ in finite $\mathbf{A}$ due to the microscopic properties of the pairs. Equation H.1 therefore becomes

$$q\mathbf{A} + n\mathbf{v} = 0. \tag{H.2}$$

Now the velocity $\mathbf{v}$ corresponds to a current density $\mathbf{J}$, where $\mathbf{J} = nq\mathbf{v}$, with n the number density of superconducting pairs. Equation H.2 therefore may be rewritten

$$\mu_0 \lambda^2 \mathbf{J} = -\mathbf{A}, \tag{H.3}$$

where $\lambda^2 = m/(n\mu_0 q^2)$; this is known as the **London equation**.

The London equation has an important consequences; taking curls of both sides, and remembering that $\nabla \times \mathbf{A} = \mathbf{B}$ and $\nabla \times \mathbf{J} = -\nabla^2 \mathbf{B}/\mu_0$ (as $\nabla \cdot \mathbf{B} = 0$) yields

$$\nabla^2 \mathbf{B} = \frac{\mathbf{B}}{\lambda^2}. \tag{H.4}$$

The implications of this equation may be most easily seen by considering a simple one-dimensional problem. Suppose a uniform magnetic field $\mathbf{B}_0$ (parallel, say to z) exists for $x \leq 0$, and that a semi-infinite block of superconductor occupies all space for $x > 0$. For such a problem, Equation H.4 becomes

$$\frac{d^2 \mathbf{B}}{dx^2} = \frac{\mathbf{B}}{\lambda^2}.$$

This has one physically meaningful solution for $x > 0$: $\mathbf{B} = \mathbf{B}_0 e^{-x/\lambda}$. In other words, the magnetic field declines exponentially as one progresses further into the superconductor. The parameter λ is known as the **penetration depth**. This is a quite distinct property of superconductors known as the **Meissner effect**; as a material enters its superconducting phase, it ejects all of the magnetic field from its interior, as long as the magnetic field is below a certain critical value.

Further reading

A good, basic summary of superconductivity which goes into more detail than I have covered here is given in Chapter 13 of *Electricity and magnetism*, by B.I. Bleaney and B. Bleaney, third edition (Oxford University Press, Oxford, 1989). More detailed treatments (which include the development of layered cuprate superconductors) are given in *Superconductivity*, by C.P. Poole, H.A. Farach and R.J. Creswick (Academic Press, San Diego, 1995) and in *Superconductivity*, by J. Annett (OUP, 2002); the former book also gives a good summary of BCS theory. It is worth examining the original classic book, *Theory of superconductivity* by J.R. (Bob) Schrieffer (Addison-Wesley, New York, 1964), which contains a great deal of enlightenment.

There is now good evidence that excitations other than phonons can provide pairing of electrons. Magnetic fluctuations have been shown to be important in heavy fermion superconductors (N.D. Mathur *et al. Nature* **394**, 39 (1998)), organic conductors (see J. Singleton, *Reports on Progress in Physics* **63**, 1111 (2000) for a summary) and in the cuprates (see J.F. Annett, *Physica C* **317–318**, 1 (1999) for an excellent and easy-to-understand review).

List of selected symbols

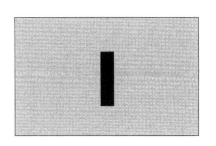

Symbol	meaning
a	atomic spacing; elsewhere edge of conventional cubic unit cell
$\mathbf{a}_j$	primitive lattice vectors in r-space
$\mathbf{A}$	magnetic vector potential; $\mathbf{B} = \nabla \times \mathbf{A}$
$\mathbf{A}_j$	primitive lattice vectors of the reciprocal lattice (in k-space)
A	cross-sectional area of Fermi surface; used as sample area in some exercises
α	αT^3 = low-T Debye contribution to heat capacity; also absorption coefficient
$\mathbf{B}, B$	magnetic flux density (also known as magnetic induction, or magnetic field)
B_m	bulk modulus
B_F	fundamental field of magnetic quantum oscillations; $B_F \equiv F$
c	speed of light
C	General heat capacity
C_{el}	electronic heat capacity
C_V	the total heat capacity of a material (per unit volume)
d	length of a sample
$d(m_1 m_2 m_3)$	separation of lattice planes with Miller indices $(m_1 m_2 m_3)$
$\mathbf{D}$	electric displacement
D_j	diffusion coefficient for carrier type j
δ	(anomalous) skin depth
Δ	energy gap of a superconductor; binding energy of an electron trap
e	the magnitude of the electronic charge; $-e$ is the charge on the electron
e	the base of natural logarithms
$\mathbf{e}_j$	unit vectors of rectangular Cartesian coordinates
E	energy (usually of a single electron)
$\mathbf{E}$	electric field (E-field)
E_A	acceptor ground-state energy
E_c	conduction-band edge energy
E_D	donor ground-state energy
E_F	the Fermi energy ($\equiv \mu(T=0)$)
E_g	energy gap of a semiconductor
E_v	valence-band edge energy
$\mathcal{E}$	E-field in one-dimensional problems
$\mathcal{E}_l$	energy of lth single-particle state
ϵ_0	permittivity of free space

ϵ_r	relative permittivity
f, f	general symbols for force (usually applied to an electron)
f	Van der Pauw scaling function
f_D	the Fermi–Dirac distribution function
F	magnetic quantum oscillation frequency; $B_F \equiv F$
$\mathcal{F}$	'spring constant' used in deriving phonon dispersion relationships
ϕ	wavefunction; also azimuthal angle of spherical polar coordinates
$g(E), g(\omega)$	density of states
$g_c(E)$	conduction-band density of states
$g_v(E)$	valence-band density of states
G	reciprocal lattice vector: $\mathbf{G} = m_1\mathbf{A}_1 + m_2\mathbf{A}_2 + m_3\mathbf{A}_3$
γ	γT is the low-temperature electronic heat capacity
h	Planck's constant
H	magnetic field strength; H-field
$\hbar$	Planck's constant over 2π
i	square root of -1; $i = (-1)^{1/2}$
i	integer or half-integer used in summations; subband index in heterostructures
I	current; also transfer integral in Exercises with explicit time dependence
j	integer or half-integer used in summations and as an index; number of dimensions
J	current density
$\mathbf{J}_q$	flux of heat (i.e. energy per second per unit area)
k_B	Boltzmann's constant
k, k	wavevector of an electron
κ	thermal conductivity
l	Landau-level quantum number; number of atoms in a basis, general integer
$L = \kappa/(\sigma T)$	Lorenz number, i.e. Wiedemann–Franz ratio/T
L_0	theoretical value of L from Sommerfeld model ($L_0 = \pi^2 k_B^2/(3e^2)$)
L_j	general sample dimension
L_z	width of semiconductor quantum well
λ	magnetic penetration depth (of a superconductor)
m	rest mass of a general particle
M	rest mass of a general particle
$\mathcal{M}$	reduced mass
M	effective mass tensor
$\mathbf{M}^{-1}$	reciprocal mass tensor
m_j	integers used in defining **G**; Miller indices of a plane
m_e	the rest mass of a free electron
m_c^*	conduction-band (band-edge) effective mass
m_{hh}^*	heavy-hole effective mass
m_{lh}^*	light-hole effective mass
μ	the chemical potential
μ^*	excitonic reduced mass

μ_c	mobility of conduction-band electrons
μ_{hh}	mobility of heavy holes
n	number of electrons or phonon states per unit volume; Bohr quantum number
n_i	intrinsic carrier density
n_j	integers used in e.g. defining $\mathbf{T}$
N	*total* number of electrons, phonons etc. in a sample
N_A	number-density of acceptors
N_c	effective number density of accessible states at the conduction band bottom
N_D	number-density of donors
N_j	number of primitive unit cells in the jth direction
N_{ph}	number of phonon modes excited at temperature T
N_s	areal carrier density
N_v	effective number density of accessible states at the valence band top
$\mathbf{p}$	momentum (e.g. of an electron)
p	number density of holes
P	hydrostatic pressure
$\mathbf{P}$	electric polarisation (dipole moment per unit volume)
$\mathcal{P}$	momentum operator
p_{N_j}	probability of occupancy of state labelled by N_j
$\mathbf{q}, q$	wavevector of a phonon or electron
q_j	charge of jth type of carrier
$\mathbf{r}$	position vector in r-space
R	resistance, short-hand for R_H in complex equations
R_H	Hall coefficient
$\rho_{xx}\ \rho_{xy}$	components of the resistivity tensor
s	sample thickness
S	entropy; also used as spin
S_z	spin component
σ	electrical conductivity
σ_{xx}, σ_{xy}	components of the conductivity tensor
t	time
t_i	transfer integral
T	temperature
$\mathbf{T}$	lattice translation vector: $\mathbf{T} = n_1\mathbf{a}_1 + n_2\mathbf{a}_2 + n_3\mathbf{a}_3$
τ^{-1}	scattering rate in the relaxation-time approximation
θ	polar angle in spherical polar coordinates
θ_D	Debye temperature
U	energy of the whole electron or phonon system
$\mathbf{v}$	velocity (e.g. of an electron)
v_x, v_y, v_x	components of velocity along the Cartesian axes
v	speed
V	typical value of electronic potential; voltage
$\mathcal{V}$	sample volume; also valence
v_ϕ	speed of sound
$V(\mathbf{r})$	potential energy (of the electron)
$V_\mathbf{G}$	coefficient of $e^{i\mathbf{G}\cdot\mathbf{r}}$ in series expansion of $V(\mathbf{r})$

V_0	average (position-independent) potential experienced by an electron in a solid
V_{kj}	k-space 'volume' in j dimensions
V_{rj}	r-space (i.e. real space) volume in j dimensions
w	width of sample
W	constant in Law of Mass-Action
x	displacement, Cartesian coordinate; also alloy composition
y, z	Cartesian coordinates

Solutions and additional hints for selected exercises

J

Chapter 1

(1.1) The quickest way to do this is to evaluate the amount of k-space required to accommodate N electrons; in 1D, this is a line of length $2k_F$, in 2D a circle, radius k_F, and in 3D a sphere, radius k_F.[1] Answers:

[1] A common mistake occurs in one dimension, when one forgets to include the states with *negative* k as well as those with positive k.

$$E_F = \frac{\hbar^2}{2m^*}\left[\frac{\pi N}{2d}\right]^2 \quad (1D)$$

$$E_F = \frac{\hbar^2}{2m^*}\left[\frac{2\pi N}{A}\right] \quad (2D)$$

$$E_F = \frac{\hbar^2}{2m^*}\left[\frac{3\pi^2 N}{\mathcal{V}}\right]^{\frac{2}{3}} \quad (3D).$$

To obtain the densities of states, use the method in Section 1.5.4; answers: $\xi = 1/2$ (1D), $\xi = 1$ (2D) and $\xi = 3/2$ (3D).

Final part: for Cu, $E_F \approx 7$ eV $\approx 81,000$ K; for the heterojunction $E_F \approx 10$ meV ≈ 116 K; for the organic solid, $E_F \approx 146$ meV ≈ 1690 K.

(1.2) (a) To get the mean electron energy U/N at $T=0$, where U is the total energy of the electron system and N is the number of electrons, it is easiest to work with k; remember $g(k) \propto k^2$. Ensuring correct normalisation we obtain

$$\frac{U}{N} = \frac{\int_0^{k_F} E(k)g(k)dk}{\int_0^{k_F} g(k)dk} = \frac{\int_0^{k_F} \frac{\hbar^2 k^2}{2m^*} k^2 dk}{\int_0^{k_F} k^2 dk},$$

leading to $U/N = \frac{3}{5}E_F$. The first law of thermodynamics is $dU = TdS - Pd\mathcal{V}$; we are at $T=0$ and N is constant, so that $P = (\partial U/\partial \mathcal{V})_N$. Now $U \propto \mathcal{V}^{-2/3}$; taking logarithms $\ln U = -\frac{2}{3}\ln \mathcal{V}$+ constants; using the trick in Section 1.5.4 you should get $P = \frac{2}{3}\frac{U}{\mathcal{V}}$. Hence $P \propto \mathcal{V}^{-5/3}$, leading to $B_m = \frac{2}{3}nE_F$.

(b) For Cu, your answer will be $B_m \approx 6.4 \times 10^{10}$ Pa; the value from the literature is 13.4×10^{10} Pa.

(1.3) (c) For Cu, your answer will be $\gamma \approx 71.1$ Jm^{-3}K^{-2} ≈ 0.503 mJmol^{-1}K^{-2}. Reason for discrepancy; hint: see part (d).

(d) Now $\gamma \propto g(E_F) \propto m^*$ (see Exercise 1.1). By taking the ratio of the γ values for Cu and UPt$_3$, you should obtain $m^* \sim 900 m_e$ in UPt$_3$.

(1.5) $(dV/dT) \approx -0.5$ μVK^{-1} at $T = 300$ K.

(1.6) Use a time-varying E-field, $\mathbf{E}(t) = \mathrm{Re}(\mathbf{E}_0 e^{-i\omega t})$. The relaxation-time equation of motion for the electron momentum $\mathbf{p}$ is then

$$\frac{d\mathbf{p}}{dt} = -\frac{\mathbf{p}}{\tau} - e\mathbf{E}_0 e^{-i\omega t},$$

which can be solved by substituting $\mathbf{p} = \mathbf{p}_0 e^{-i\omega t}$. Remembering that the current density $\mathbf{J} = -ne\mathbf{p}/m_e$,

$$\mathbf{J} = \frac{ne^2\tau}{m_e} \frac{1}{(1-i\omega\tau)} \mathbf{E},$$

so that the conductivity σ is

$$\sigma = \frac{ne^2\tau}{m_e} \frac{1}{(1-i\omega\tau)}.$$

This becomes the Drude result as $\omega \to 0$.

To get ϵ_r, we need the polarisation $\mathbf{P}$, which is obtained from the displacement $\mathbf{x}$ of the electron:

$$\mathbf{x} = \int \frac{\mathbf{p}}{m_e} dt = \frac{e\tau}{i\omega m_e} \frac{\mathbf{E}_0 e^{-i\omega t}}{(1-i\omega\tau)}.$$

The equation $\mathbf{P} = -ne\mathbf{x}$ combined with $\mathbf{D} = \epsilon_r \epsilon_0 \mathbf{E} = \epsilon_0 \mathbf{E} + \mathbf{P}$ is then sufficient to obtain ϵ_r. In the limit $\omega\tau \gg 1$,

$$\epsilon_r = 1 - \frac{ne^2}{\epsilon_0 m_e \omega^2} = 1 - \frac{\omega_P^2}{\omega^2},$$

where ω_P is the plasma frequency. This implies that the refractive index ($\equiv \epsilon_r^{\frac{1}{2}}$) becomes totally imaginary when $\omega < \omega_P$. Hence the only solution for electromagnetic radiation within the metal will be evanescent (decaying) waves,[2] so that the material will be 100% reflecting. For $\omega > \omega_P$, the material will transmit radiation.[3] For Cu, $\omega_P \approx 1.6 \times 10^{16}$ radians s^{-1}, corresponding to 2.6×10^{15} Hz, or around 10 eV.

[2] See e.g. page 235 of *Electricity and magnetism*, by B.I. Bleaney and B. Bleaney, third edition (Oxford University Press, Oxford, 1989).

[3] The transparency of metals in the ultraviolet is discussed on page 18 of *Solid state physics*, by N.W Ashcroft and N.D. Mermin (Holt, Rinehart and Winston, New York, 1976), along with plasma oscillations.

Chapter 2

(2.2) Hint: use the equation for a plane $\mathbf{r}.\mathbf{n} = d$, where $\mathbf{n}$ is a unit vector normal to the plane and d is the perpendicular distance from the origin to the plane. Propose that $\mathbf{n} = \mathbf{G}/|\mathbf{G}|$ and try three suitable values for $\mathbf{r}$ lying on the plane (Hint: see Fig. A6; at what points does the plane cut the axes?) to obtain d.

(2.3) The dispersion curves have the general form

$$E = \frac{\hbar^2}{2m_e}(\mathbf{k} - \mathbf{G})^2.$$

For the square lattice

$$\mathbf{G} = \left(\frac{2\pi}{a}\mathbf{i}, \frac{2\pi}{a}\mathbf{j}\right),$$

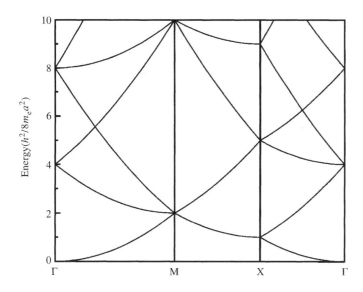

Fig. J.1 Solution to Exercise 2.3.

where i, j are $0, \pm 1, \pm 2 \dots$. Thus, along Γ–M, the dispersion curves look like

$$E = \frac{\hbar^2}{2m_e}\left(\frac{\pi}{a}\right)^2 \left[(\delta - 2i)^2 + (\delta - 2j)^2\right]$$

where δ runs from 0 (Γ) to 1 (M). Along M–X, the dispersion curves look like

$$E = \frac{\hbar^2}{2m_e}\left(\frac{\pi}{a}\right)^2 \left[(1 - 2i)^2 + (\delta - 2j)^2\right]$$

where δ runs from 1 (M) to 0 (X). Finally, along X–Γ, the dispersion curves look like

$$E = \frac{\hbar^2}{2m_e}\left(\frac{\pi}{a}\right)^2 \left[(\delta - 2i)^2 + (-2j)^2\right]$$

where δ runs from 1 (X) to 0(Γ). The solution is shown in Fig. J.1.[4]

(2.4) For bcc $|\mathbf{A}_j| = 2\sqrt{2}(\pi/a)$; for fcc $|\mathbf{A}_j| = 2\sqrt{3}(\pi/a)$

Chapter 3

(3.2) (a) The simplest way to obtain the gaps is to look at the difference in the expectation values of the energy for the two wavefunctions, $E_{\cos} - E_{\sin}$. The only term in the Hamiltonian that affects the two wavefunctions differently is the potential term; therefore we end up evaluating

$$\int_{-a/2}^{a/2} E_{\cos} \cos^2(\frac{\pi x}{a}) dx = \text{constants} + \int_{-a/2}^{a/2} v \cos(\frac{2\pi x}{a}) \cos^2(\frac{\pi x}{a}) dx)$$

and

$$\int_{-a/2}^{a/2} E_{\sin} \sin^2(\frac{\pi x}{a}) dx = \text{constants} + \int_{-a/2}^{a/2} v \cos(\frac{2\pi x}{a}) \sin^2(\frac{\pi x}{a}) dx,$$

[4] Free-electron band calculations such as these are sometimes referred to as 'empty lattice band calculations'. If the appropriate crystal symmetry (i.e. reciprocal lattice) is used, the empty lattice bands often give a reasonably accurate representation of the bandstructure of real solids at energies well above the occupied bands. A good example of this is given by G.P Williams *et al.* (*Phys. Rev. B* **34**, 5548 (1986)), who used ARPES (see Chapter 8) to study the high energy bands of InSb, GaAs, GaP, GaSb, InP and InAs.

which gives $E_{\cos} - E_{\sin} = \nu$.

(b) The metal–insulator transition occurs when $\nu = \hbar^2 \pi^2/2m_e a^2$, i.e. as the lowest point of the upper band goes above the highest point of the lower band; see Fig. 3.4.

(3.3) (b) Assume that the necks are a small perturbation of the Fermi surface, so that $E_F \approx$ the free-electron value for Cu, i.e. 7 eV (see Exercise 1.1). We then assume that the energy at the top of the lower band is given by

$$E \approx \frac{\hbar^2 k^2}{2m_e} - \frac{E_g}{2}$$

on the Brillouin-zone boundary, where $E_g = 4$ eV is the energy gap. Points k' (measured from the zone centre) on the neck are therefore given by

$$E_F \approx \frac{\hbar^2 k'^2}{2m_e} - \frac{E_g}{2}.$$

Now the point on the Brillouin-zone boundary on the axis (centre) of the neck is at a distance $k = \sqrt{3}\pi/a$ from the zone centre; by Pythagoras's theorem

$$k'^2 = \left[\frac{\sqrt{3}\pi}{a}\right]^2 + k_n^2,$$

where k_n is the neck radius. This yields $k_n \approx 3.37 \times 10^9$ m^{-1} (cf. $k_F \approx 13.6 \times 10^9$ m^{-1}). The estimate is not bad; in reality accurate measurements give $k_n/k_F \approx 1/7$.

(3.4) (b) The approximations involved are similar to those in Exercise 3.3. We assume that the band gaps at the zone boundary do not shift E_F much from the free-electron value, $E_F = \frac{\hbar^2}{2m_e}\left[\frac{2\pi}{a^2}\right]$, where a is the edge of the square unit cell. The Fermi surface just touches the Brillouin-zone boundary when

$$E_F \approx \frac{\hbar^2 \pi^2}{2m_e a^2} - \frac{E_g}{2},$$

where $\frac{\hbar^2 \pi^2}{2m_e a^2}$ is the free-electron energy at the points on the zone boundary closest to the zone centre. This yields the energy of feature in optical absorption $= E_g = \frac{\hbar^2}{m_e a^2}(\pi^2 - 2\pi) \approx 3.0$ eV.

Chapter 5

(5.1) See Section 9.1.

(5.3) (b) From Chapter 1, you will remember that $C_{el} \propto g(E_F) \propto n/E_F$; virtually all of the other details of the bands (e.g. the degeneracy) are killed off in the differentiation used to derive eqn 1.30, as long as the bands are reasonably simple. Assume that the electron densities are similar in Cu and Sc. Answer: $C_{el}(\text{Sc})/C_{el}(\text{Cu}) \approx 7.5$.

(5.4) (a) $\mathbf{G} = \frac{2\pi}{a}\mathbf{e}_1$ in this case. For motion parallel to the x direction with $\mathbf{k}$ close to the zone boundary, we consider $\mathbf{k} = (\frac{\pi}{a} + \Delta k_x)\mathbf{e}_1$ and $\mathbf{k} - \mathbf{G} =$

$-(\frac{\pi}{a} - \Delta k_x)\mathbf{e}_1$, where Δk_x is small. Substituting this into the equation for the energy, one obtains after a bit of fiddling

$$E = \frac{\hbar^2}{2m_e}\frac{\pi^2}{a^2} + \frac{\hbar^2}{2m_e}\Delta k_x^2 \pm \left[(\frac{\hbar^2}{2m_e}\Delta k_x \frac{2\pi}{a})^2 + |V_\mathbf{G}|^2\right]^{\frac{1}{2}}.$$

We then make Δk_x very small, such that $(\frac{\hbar^2}{2m_e}\Delta k_x \frac{2\pi}{a}) \ll |V_\mathbf{G}|$; this allows several approximations, including the expansion of the square root, to be made, so that

$$E \approx \frac{\hbar^2}{2m_e}\frac{\pi^2}{a^2} \pm \left[|V_\mathbf{G}| + 2\frac{\frac{\hbar^2}{2m_e}\frac{\pi^2}{a^2}\frac{\hbar^2}{2m_e}\Delta k_x^2}{|V_\mathbf{G}|}\right].$$

For the first part we require $(\mathbf{M}^{-1})_{11}$, as the motion is in the x ($\mathbf{e}_1$) direction. Remembering that $(\partial^2 E/\partial k_x^2) \equiv (\partial^2 E/\partial \Delta k_x^2)$,

$$(\mathbf{M}^{-1})_{11} = \frac{1}{\hbar^2}\frac{\partial^2 E}{\partial \Delta k_x^2}.$$

The required effective mass is then

$$((\mathbf{M}^{-1})_{11})^{-1} \approx \pm m_e \frac{V_\mathbf{G}}{2\frac{\hbar^2}{2m_e}\frac{\pi^2}{a^2}}.$$

The $+$ sign corresponds to the bottom of the upper band (electron-like), and the $-$ sign (hole-like) to the top of the lower band.
For motion parallel to $\mathbf{e}_2$, use $\mathbf{k} = \frac{\pi}{a}\mathbf{e}_1 + k_y \mathbf{e}_2$ and $\mathbf{k} - \mathbf{G} = -\frac{\pi}{a}\mathbf{e}_1 + k_y \mathbf{e}_2$. The resulting expression for the energy is much simpler, so that it is relatively easy to obtain

$$((\mathbf{M}^{-1})_{22})^{-1} = (\frac{1}{\hbar^2}\frac{\partial^2 E}{\partial k_y^2})^{-1} = m_e.$$

(5.5) Almost empty: $m^* = \hbar^2/2t_1 a^2$; almost full: $m^* = -\hbar^2/2t_1 a^2$.

Chapter 6

(6.1) The temperature dependence of the conductivity will be dominated by the exponential part of the temperature dependence of the intrinsic carrier density

$$n_i \propto \exp(-E_g/2k_B T).$$

Comparing the conductivities σ_1 and σ_2 for two closely-spaced temperatures T_1 and T_2 we obtain

$$\frac{\sigma_2}{\sigma_1} \approx \frac{n_{i2}}{n_{i1}} \approx \frac{\exp(-E_g/2k_B T_2)}{\exp(-E_g/2k_B T_1)} = \exp\left[-\frac{E_g}{2k_B}(\frac{1}{T_2} - \frac{1}{T_1})\right].$$

Setting $T_1 = 300$ K and $\frac{\sigma_2}{\sigma_1} = 1.1$ we obtain a temperature increase ≈ 0.5 K.

(6.2) $n_i \sim 10^{13}$ cm^{-3}

(6.4) We must first find the density of Si atoms in a sample of Si; using Fig. 4.4 and Table 4.1, this is $\approx 5.00 \times 10^{28}$ m^{-3}. If one part in 10^9 are replaced by donors, then $N_D \approx 5.00 \times 10^{19}$ m^{-3}. Estimates of n_i using eqn 6.14 yield $n_i \approx 6 \times 10^{15}$ m^{-3}, i.e. $N_D \gg n_i$. Assuming all the donors are ionised ($n \approx N_D$), the chemical potential can then be estimated using eqn 6.11, to give values $\sim 200\text{–}300$ meV below the conduction-band edge.

(6.5) By counting conduction-band states, you should obtain

$$\frac{\hbar^2}{2m_c^*}(3\pi^2 n)^{\frac{2}{3}} = xk_B T.$$

Rearrange to get an expression for n, and equate this to N_c. All of the constants should cancel, leaving $x \approx 1.2 \sim 1$; in other words, N_c is a useful guide to the number of accessible states at the band edge.

Chapter 7

(7.2) The solutions are given in Table J.1.

Table J.1 Subband energies (in units meV) for the conduction-band (electron) and valence-band (light- and heavy hole) subbands in the GaAs-(Ga,Al)As quantum well of Exercise 7.2. The $i = 3$ subband of the light holes is marked with * because it is only just bound; depending on the accuracy of your calculation, it may or may not show up.

i	electron	light hole	heavy hole
1	30.6	23.9	6.3
2	120	93	25.5
3	250	180*	56.8
4	–	–	99.6
5	–	–	151

(7.3) The allowed transition energies are given in Table J.2. In the absence of exctionic effects, the absorption will show a step upwards at each transition, and be flat elsewhere.

Chapter 8

(8.1) (a) Use eqn 8.38, and assume that the lowest minimum corresponds to $l = 0$. The energy spacing of the minima is $\frac{\hbar eB}{m_{\text{hCR}}^*} + \frac{\hbar eB}{m_{\text{cCR}}^*}$. Using these two facts $E_g \approx 0.42$ eV.
(b) Using the binding energy of the donor and the dielectric constant of the semiconductor leads to $m_{\text{cCR}}^* \approx 0.036 m_e$. Hence, using the spacing of the minima in Fig. 8.29, $m_{\text{hCR}}^* \approx 0.5 m_e$.

Table J.2 Transitions allowed under the selection rule $\Delta i = 0$; lh = light hole, hh = heavy hole, e = electron. *The $i = 3$ subband of the light holes is only just bound; depending on the accuracy of your calculation, it may or may not show up.

transition	energy (eV)	transition	energy (eV)
hh1-e1	1.557	lh1-e1	1.574
hh2-e2	1.666	lh2-e2	1.733
hh3-e3	1.827	lh3-e3*	1.950

(8.5) The metal's Fermi surface is extended in the **c** direction. It has extremal 'necks' of cross-sectional area 1.26×10^{19} m^{-2} and 'bellies' of cross-sectional area 2.71×10^{20} m^{-2}. The metal is monovalent, since the volume of the Fermi surface appears to be around half that of the Brillouin zone.

(8.6) The cyclotron mass is $0.37 m_e$. The field must be aligned to better than $\sim \pm 0.5°$ to stop the helical electron orbits from drifting out of the skin depth before they have received a few 'kicks' from the GHz E-field.

(8.7) The masses are $0.1 m_e$ and $0.6 m_e$. The mobilities are 10 m^{2}V^{-1}s^{-1} and 2 m^{2}V^{-1}s^{-1}.

(8.8) Ratio of neck to belly radius $\approx 1/7.2$.

(8.9) $B_F = F = 27\,300$ T; $n = 2.55 \times 10^{28}$m^{-3}; $E_F = 2.43$ eV.
Estimation of actual bulk modulus: the de Haas–van Alphen frequency $F \propto k_F^2$ for a spherical Fermi surface. Hence $F \propto \mathcal{V}^{-2/3}$, where $\mathcal{V}$ is the sample volume. The bulk modulus is $B_m = -\mathcal{V}(\partial P/\partial \mathcal{V})_T$; integrating, we obtain $B_m \ln(\mathcal{V}_2/\mathcal{V}_1) = -P$, where $\mathcal{V}_1$ is the volume at $P = 0$ and $\mathcal{V}_2$ is the volume at pressure P. Hence $(F_2/F_1) = \exp(2P/3B_m)$, where F_1 is the de Haas–van Alphen frequency at $P = 0$ and F_2 is that at finite pressure P. For $P \ll B_m$, $(F_2/F_1) \approx 1 + (2P/3B_m)$. Hence the graph can be used to give the actual bulk modulus: $B_{\text{act}} \approx 5.7 \times 10^9$ Pa. The bulk modulus calculated using the Sommerfeld model (see Exercise 1.2) is 6.6×10^9 Pa.

Chapter 9

(9.1) (a) At $T = 30$ K, the phonon dispersion relationship can be approximated by $\omega = v_\phi k$ (see Appendix D), with v_ϕ the velocity of sound; $v_\phi = (B/\rho)^{\frac{1}{2}}$, with B the bulk modulus and ρ the density. A typical phonon wavevector is $q \sim k_B T/\hbar v_\phi$; the Fermi wavevector of Na can be derived from the density and atomic mass, leading to a scattering angle $\sim 10°$.
(b) At $T = 300$ K, a typical phonon has a q close to the Brillouin-zone boundary (as $T > \theta_D$). Typical scattering angles are $\geq \sim 70°$.

(9.2) For both phonons and electrons,
$$\kappa \approx \frac{1}{3} C_V <v> \lambda,$$

where C_V is the heat capacity *per unit volume* (due to phonons or electrons, respectively), $<v>$ is the average velocity of the phonon or electron and λ is its mean free path length. For phonons $\lambda \sim$ size of the wire (see last part of question), and C_V is given by the Debye model (see Appendix D). For the phonon velocity, see the method of the previous question. (NB, there is a short cut to the electronic thermal conductivity from the data given!) Answer: ratio of thermal conductivity due to phonons to that due to electrons is $\sim 10^{-3}$.

(9.3) Force per unit volume on impurities $= 8.2 \times 10^7$ Nm^{-3}.

(9.5) The scattering cross-section for Rutherford scattering is proportional to Z_{eff}^2, where Z_{eff} is the effective charge of the ion. For Ga^{3+}, $\Delta\rho \approx 1.8 \times 10^{-9}$ Ωm. For vacancies, $\Delta\rho \approx 0.44 \times 10^{-9}$ Ωm.

(9.8) $\Delta \approx 0.69$ meV.

Chapter 10

(10.1) (i) Acceptors, with $N_A \approx 3.1 \times 10^{21}$ m^{-3}. (ii) $E_g \sim 0.25 - 0.35$ eV, depending on the exact region of the plot used to obtain the gradient.

Chapter 11

(11.1) (b) From the periodicity of the oscillations in $1/B$, $N_s \approx 2.7 \times 10^{15}$ m^{-2}. Mean free path length ≈ 0.2 μm. (During the derivation of the mean free path length, you will find that it is equal to $\frac{h}{e^2}\frac{1}{\rho_{xx}}\frac{1}{k_F}$; strange but true!)[5]

(c) 0.5 meV.

[5] The first term is the inverse of the quantum of conductance. (See also pages 148 and 162–163.)

(11.2) Two subbands are occupied; from the frequencies of the oscillations they have areal carrier densities 4.61×10^{15} m^{-2} and 0.12×10^{15} m^{-2}, giving a total of 4.73×10^{15} m^{-2}.

Chapter 12

(12.3) $\mu_h \approx 0.2$ m^{2}V^{-1}s^{-1}; $D_h \approx 0.008$ m^{2}s^{-1}.

(12.7) Typical energy level spacing $\sim$ a few tenths of an meV.

Index

Note. In the case that several pages are given for a particular topic, the most important is/are shown in **bold**.

acceptors 59–60
Ag (Silver) 116
(Al,In)As (Aluminium Indium Arsenide) 66, 71, 152
alkali metals
  Fermi surfaces 27
  resistivity of 126
  thermal conductivity of 126
  tight-binding model of 36
Allen–Bradley resistor 129
alloys, semiconductor 65–6
angle-resolved photoelectron spectroscopy **103–4**, 106, 114, 211
anion molecule 76
anomalous skin depth, *see* skin depth 98
ARPES, *see* angle-resolved photoelectron spectroscopy
atomic fine structure constant 150
avalanching 160
averaging postulate 175
Azbel'–Kaner geometry 98

ballistic transport 161
band, electronic
  dynamical properties in absence of scattering 117–19
  filling 20, 26
  general properties of 41–7
  introduction 20
  nearly-free electron model 26–30
  tight-binding model 32–40
  thermal population of, 56–9
band gap 20, 49
  alloys 66
  direct 51, 53–4
  in the nearly-free electron model 25, 26
  in the tight-binding model 36, 37
  indirect 51
bandstructure, electronic
  basic ideas 20
  engineering 65–81
  general properties of 41–7

measurement of 85–105
nearly-free electron model 26–9
tight-binding model 34–9
bandwidth 35
Bardeen–Cooper–Schrieffer theory 202–4
barrier (in a quantum well) 68
basis 165
  and phonons 184–5
bcc, *see* body-centred cubic
BCS, *see* Bardeen–Cooper–Schrieffer theory
BEDT-TTF, *see* bisethylenedithiotetrathiofulvalene
Bi (bismuth) 135
Bible, quotation from vii
binary semiconductors (*see also* III–V and II–VI semiconductors)
  Brillouin zone 37
  crystal structure 37
bisethylenedithiotetrathiofulvalene
  compounds of 75–8, 96
  molecule 75
bistability 160
Bloch oscillations 118
Bloch's theorem
  definition 20, 32
  derivation using plane waves **16–20**
  derivation, alternative 21
Bloch wavefunctions
  envelope function approximation 74–5
  general properties 32, **44–5**
  scattering 45
body-centred cubic 167
Bohr model 191–3
  excitons 192–3
  hydrogen 191
  hydrogenic impurities 192
Bohr radius (of an impurity) 60, 191
Boltzmann transport equation 130–1
Born–von Karman boundary conditions 17, 18
Bose–Einstein distribution function 178

Bosons 178
Bragg condition 26
Bragg diffraction, condition for 30, 31
Bravais lattice 171
Brillouin zone
  definition 17, 167
  number of k-states within 18, 44
  of square, bcc and fcc lattices 21–2
  volume 18
buffer layer 68
bulk modulus
  electronic 8, 14
  of a metal 8, 14

C (carbon) 36
canonical momentum 200
carrier
  as general term for electrons or holes 8, 56, 58, 60, 133, 154
  injection 154–5
  minority and majority 154
  recombination and extraction 155
carrier density
  semiconductors, extrinsic 59–62
  semiconductors, intrinsic 58–9
  typical metallic 4, 8
CdTe
  Brillouin zone and crystal structure 37
  electroreflectance spectrum 105
  phonons 183, 189
charge-density wave (CDW) 82
charge-transfer salts, *see* organic metals
chemical potential 9, 12, 176
  in semiconductors 58, 62
  in the quantum Hall effect 149–50
closed orbits 139
compensating coil 94
compensated semimetal 135
compensation 60
composite fermions 151
conductance, quantum of 148, 162–3, 216
conduction band

218 *Index*

definition 49
effective masses 54
conductivity (*see also* electrical *or* thermal conductivity)
   measurement of 194–9
   tensor 136–8, 145–50
conventional versus primitive 169–71
convolution 165
Corbino geometry 199
Coulomb pseudopotential 202–3
critical current 202
critical field 202
critical temperature 202
cryogenic techniques 114, 199
crystal 165–6
   designation of directions within 167–8
   designation of planes 168–9
crystal momentum 42
Cu (copper)
   bandstructure 37–8, 48
   Brillouin zone 39
   electrical and thermal conductivity of 126
   Fermi surface measurement 105–6, 139, 142
   'necks' 30, 38, 39, 92–3
   thermodynamics of 14, 15
cubic-F lattice 167
cubic-I lattice 167
cyclotron frequency
   for orbits about Fermi surface 85–7
   quantum-mechanical derivation 87–90
   semiclassical derivations 47, 85–6
cyclotron mass 86, 89
cyclotron motion 85–7
   electrons versus holes 87
cyclotron resonance 97
   in metals 98, 106, 115
   in semiconductors 98–100, 115
   selection rules 99
   semiclassical derivation using effective mass tensor 47

DBRTS, *see* double-barrier resonant tunnelling structure
de Haas-van Alphen effect/oscillations **94–5**, 105–6, 115–16
Debye heat capacity 123, **185–6**
Debye model of phonons 185–90
   assumptions 185–6
   comparison with 'real' phonons 186

Debye temperature 123–5, **188–9**
   and superconductivity 203
   of common elements and compounds 189
deep traps 62–3
degeneracy, of donor states 61
degenerate two dimensional electron systems 73–4
degenerate semiconductor 62
delocalised states, *see* extended states
δ-doping layer 74
density of states 172–4
   and effective mass 14, 43–4
   band electrons in a magnetic field 91
   dimensionality 14, 69, **172–3**
   electronic 9–10, 14
   heavy hole 51
   one dimensional 161–3
   quantum well 69–70
   semiconductors 55, 57
   two-dimensional quantum Hall system 149
diamond 36
diamond structure 36
diffusion 155–6
   coefficient 155, 157
dimensionality
   effect of magnetic field 91, 143–4, 149
   effect on electronic density of states 14, **172–4**
   effect on phonon density of states 189–90
   of Fermi surfaces 78
   oxides 80–1
   quantum point contacts 161–2
direct gap (*see also* II–VI and III–V semiconductors) 53–4
disorder and quantum Hall effect 148–50
disordered systems, electrical resistivity 129–31
dispersion relationship
   band electrons 44
   definition 20
   periodicity in k-space 17, 24, 26
   quasiparticle 113
distribution functions, derivation of 175–9
divalent metals 27
   two dimensional 28, 29
dog's bone 93
donors 59–61

double-barrier resonant tunnelling structure 160–1
double exchange 79
drift velocity 120–1, 156
   electric field dependence 158–9
   saturation 158
Drude model of metals 2–7
   assumptions 2
   disclaimers 2, 7
   electrical conductivity 4
   failures of 5–7, 10
   heat capacity 4
   thermal conductivity 4
dynamics of band electrons 44, 47

effective filling factor 151
effective g-factor 148
effective mass
   and density of states 14, **43–4**
   cyclotron 86, 89
   derivation 42–3
   different types of, 113
   effect of interactions 113
   gentle introduction to 14
   II–VI and III–V semiconductors 54
   infinite 118
   measurement using cyclotron resonance 97–100
   measurement using interband magneto-optics 100–2
   measurement using magnetic quantum oscillations **96–7**, 107–8
   oscillatory 117–19
   role in magnetoresistance 138
   tensor **47**, 88–9, 99
   tight-binding model 35
Einstein equation 156
elastic process 122
electrical conductivity
   disordered systems 129–31
   metals 119–27
   relaxation-time approximation 4
   semiconductors 127–9
   Sommerfeld model 14
   tensor 136–8, 145–50
   thin films and wires 131
electron density, *see* carrier density
electron–electron interactions 112–14
electron–electron scattering 125–7
electron–phonon interactions 112–14
electron–phonon scattering 122–7, 129
electronic bandstructure, *see* bandstructure

electronic density of states 9–11
electronic potential, *see* potential energy
electroreflectance 104–5
empty lattice band calculations 211
energy gap, in semiconductors, *see* band gap
energy gap, in superconductors 201–3
envelope-function approximation **74–5**, 83–4
epitaxy 66–7
excitons **54–6**, 69–70, 102
    energy levels 192–3
    extraction of binding energy 102
    reduced mass 192–3
exchange interactions 79
excuses, feeble 8
extraction 155
extended states 148
extremal orbit/cross-section 92
extrinsic carrier density 59–62

face-centred cubic 167
fan diagram 101–2
far-infrared laser 98
fcc, *see* face-centred cubic
Fermi energy
    chemical potential versus 9, 12
    definition 9
    dimensionality 14
    Sommerfeld model 8
    typical metallic values 8
Fermi–Dirac distribution function 9
    derivation 178
    effect on magnetic quantum oscillations **96–7**, 150
    in semiconductor heterojunctions 145, 150
    in semiconductors 56–62
    typical form in metals 9-10, 15
Fermi–Dirac integrals 11
Fermi gas 113
Fermi level 9
Fermi liquid 113
Fermi surface
    and electrical/thermal conductivity 120–2
    construction of 29
    cyclotron motion about 86–7, 139
    dimensionality of 78
    extremal orbit/cross-section 92
    measurement of 94–5, 105–6,138–9
    nearly-free electron model 27–31

organic metal 77–8
    Sommerfeld model 8
    two-dimensional semiconductor system 73, 151
    velocity at 78, 120
    warping 143
Fermi velocity 8, 78, 120
Fermi wavevector
    free-electron 8
    typical metallic values 8
fermions 173, 178
    composite, *see* composite fermions
    heavy, *see* heavy fermion
filling factor 148–51
flux quantum 151
force, effect on band electron **42–3**, 85, 117–9
four-wire method 194–5
Fourier analysis, of magnetic quantum oscillations 95
fractional quantum Hall effect 150–2
Franz–Keldysh effect 104
free-electron approximation 2
freeze-out 61
function of state, thermodynamic 94
fundamental field 92

g-factor 97, 148
(Ga,Al)As (Gallium Aluminium Arsenide)
    applications 65–6, 68–71
GaAs (Gallium Arsenide)
    bandstructure 53–5
    Brillouin zone and crystal structure 37, 49
    constant energy surfaces 56
    drift velocity 158–160
    effect of pressure on 141–2
    Gunn effect 158–60
    intrinsic carrier density 59
    heterojunction 73–4, 143–52
    optical properties 54–5, 101–2
    phonons 184, 189
    quantum point contact 161–3
    quantum wells and superlattices 68–71
(Ga, In)As (Gallium Indium Arsenide) 66, 71–2, 152
gate 161
Ge (Germanium)
    band gap 36, 37, 50–1
    bandstructure 50–1

Brillouin zone and crystal structure 37, 49
    constant energy surfaces 52
    effective masses 100
    hydrogenic donors 60
    intrinsic carrier density 59
    optical absorption 51–2
    tight-binding model of 36–7
glamour of condensed matter physics v
grand partition function 175–7
group IA metals, *see* alkali metals
group IIA metals 27, 36
group IV elements
    band gaps 36, 37
    Brillouin zone and crystal structure 37, 49
Gunn diode 160
Gunn effect 158–60

Hall coefficient 6, 135, 139, 146–7
Hall effect
    experimental details 194–9
    experimental geometry 134
    majority carrier 158
    metals, typical data 7
    more than one type of carrier 133–5
    single carrier species 6
    two-dimensional, *see* quantum Hall effect
    van der Pauw method 197–8
Hall plateau 147–50
heat bath 175
heat capacity
    classical gas 178–9
    electronic, Drude result 4
    electronic, Sommerfeld result 11–12
    metallic, typical data 5, 14
heating during measurements 195
heavy-fermion compounds 14, **113–14**
heavy hole
    definition 51
    density of states 51
    effective masses 54
    mobility 128
HEMT, *see* high electron mobility transistor
heterojunction 73–4, 99
(Hg,Cd)Te
    applications 65–6
    band gap 66
(Hg,Cd)Te *cont.d*
    Brillouin zone and crystal structure 49

photoconductivity 63
superlattices 71
high electron mobility transistor 74
high T_c cuprate superconductors 80–1, 83
holes
  cyclotron motion of 87
  density, *see* carrier density
  derivation of properties 45–6
  dynamics of 118
  heavy, *see* heavy hole
  light, *see* light hole
  low mobility of, 128
hopping
  thermally-activated 129
  variable range 130
horizontal process 122
hot carrier effects 158–61
hot topics vii
hydrogenic donors, impurities 60, 192

III–V and II–VI semiconductors
  bandstructure 53–5
  Brillouin zone and crystal structure 37, 49
  constant energy surfaces 56
  optical properties 54–5, 63
impurities
  acceptors and donors 59–60
  energy levels 192
  hydrogenic 192
  ionisation 61, 63, 64
  scattering of electrons by, 45, 125, 128
impurity bands 62
InAs (Indium Arsenide) 54, 66
InAs-(Ga,In)Sb superlattices 72–3
independent electron approximation 2
indirect-gap 51
inelastic process 122
injection 154
InP (Indium Phosphide) 66, 71
InSb (Indium Antimonide)
  band gap and effective masses 54
  magnetophonon effect 140
  photoconductivity 63
insulators
  caused by absence of scattering 119
  definition 49
  nearly-free electron model 27, 30
  versus semiconductors 62–3
integer/integral quantum Hall effect, *see* quantum Hall effect

interband magneto-optics 100–2, 115
intrinsic carrier density 58–9
ionic cores 2, 8
ionisation of impurities 61, 63, 64
isotope effect 201–3

joint band diagram 105
joint density of states 105
joke 86
junction transistor 158

k (wavevector)
  as a quantum number 42
  basic definition 16
K (potassium), *see* alkali metals
k-space
  periodicity of 17, 24
Knudsen cells 67
k.p method 64

Landau Fermi liquid theory 112–14
Landau gauge 88
Landau level filling factor 148
Landau levels 87–92, 143–8, 150
Landau quantisation 85, 89, 90
Landau tubes 89–91
lattice 165
  translation vectors 16, **165–6**
Laughlin wave function 152
law of mass-action **56–8**, 127
layered conducting oxides 78–81
layered materials, measurement of resistivity tensor 198–9
light hole
  definition 51
  effective masses 54, 100
localised states 148
lock-in amplifier 105
longitudinal optic phonons 139–41
Lorentz force
  effect on band electrons 85–7
  Hall effect 133–5, 147
Lorenz number, *see* Wiedemann–Franz ratio

macrostate 175
magnetic breakdown 97, 108, 110
magnetic quantum oscillations **91–7**, 107–10, 114
magnetic vector potential 200
magnetisation 94–5

magnetophonon effect/resonance 139–41
magnetoresistance
  determination of Fermi-surface shape using 138–9
  due to modification of carrier motion 133–9
  in magnetic oxides 78–80
  in two-dimensional systems 143–52
  longitudinal 136
  transverse 136
majority carriers 154, 158
malleability of metals 1
Matthiessen's rule 122
Maxwell–Boltzmann distribution 10, 178–80
MBE, *see* molecular beam epitaxy
Meissner effect 203–4
metals
  alkali 27, 36, 126
  bonds 1
  bulk moduli 8, 14
  carrier densities 4, 8
  divalent 27, 36
  Drude model of 2–7
  early history of research 1, 7
  electrical conductivity 4, **119–27**
  general properties 1–2
  group IIA 27, 36
  Hall effect 7
  heat capacity 5
  magnetoresistance 133–9
  malleability 1
  nearly-free electron model of 23–30
  one-dimensional 27, 81–3
  plasma frequency 15
  quantum statistics of 7–13
  Sommerfeld model of 7–13
  thermal conductivity 5, **119–27**
  transition 37–9, 126–7
metal–insulator transition
  disordered systems 131
  nearly-free electron model 28, 30
  Peierls transition 81–3
metal–organic vapour-phase epitaxy 66–7
microstate 175
Miller indices 21, 31, **168–9**
minority carriers 154, 156–8
  excitation of, 154–5, 157
mixed valence 79
mobility 128, 156, 158
  measurement of 194–9
  spectrum analysis 198

MOCVD, *see* metal–organic vapour-phase epitaxy
modulation coil 94
modulation doping 73
modulation spectroscopy 104
molecular beam epitaxy 66–7
monovalent metals 27, 81
MOVPE, *see* metal–organic vapour-phase epitaxy
multilayers 65–73

Na (Sodium), *see* alkali metals
NDC, *see* negative differential conductance
nearly-free electron model 23–30
   band gaps within 25, 26
   Bragg condition 26
   comparison with tight-binding model 41
   metal-insulator transition 28, 30
   qualitative explanation of properties of real elements 27–9
neck and belly orbits 92–3
'necks' in Cu 30
negative differential conductance 160
nesting vector 82
new effects, cautionary remarks 195
nonparabolic bands 63
nonparabolicity, measurement of 102

OMVPE, *see* metal-organic vapour-phase epitaxy
open orbits 93, 139
optic phonons 139–40, 158, **183**
optical absorption
   quantum wells 69–70, 83–4
   semiconductors 51–2, 54–5, 63
optoelectronics 70
orbits, about the Fermi surface 85–7
   extremal 92
   open 93
   quantisation of 90–1, 93
organic metals 75–8, 83
   bandstructure 76, 78
   Fermi surfaces 77–8
   magnetic quantum oscillations 96, 106–11
   superconductivity 76–7
oxides 78–81, 83

parabolic bands 43

partially-filled bands 76
Pauli's exclusion principle 173
PbTe (Lead Telluride) 101–2
Peierls transition 81–3
periodic boundary conditions 17, 172
perovskite structure 80
phase-sensitive detection 105
phase space 7, 14
phonon(s) 181–90
   absorption 122, 139–41
   acoustic 183
   and effective mass 113
   'ball and spring' model 182–4
   'black-body' distribution 188
   Brillouin zone and 183
   characteristic energy of 123, **188**
   Debye model 185–90
   Debye temperature 188–90
   definition 181
   dimensionality 189–90
   emission 122, 158, 160
   involvement in optical absorption of 51–2
   involvement in superconductivity of 201–3
   longitudinal 184
   magnetophonon effect 139–41
   number 187–8
   optic 139–40, 158, **183**
   relationship to molecular vibrations 184–5
   scattering of electrons by 45, **122–9**, 139–41, 158
   three dimensional 184–5
   transverse 184
   versus normal modes 181
photoconductivity 63, 154
plane waves 18, 25
plasma frequency 15
plasma oscillations 210
plateau, in quantum Hall effect 147–50
pressure, effect on bandstructure 109–11, 116
primitive lattice translation vector, unit cell 165–7
   versus conventional 169–71
polaron 112
potential energy, electronic
   as a Fourier series 16, 18
   Drude and Sommerfeld models 3, 8
   commensurability with electronic wavefunctions 25
   tight binding model 32, 33

translational symmetry of, 16–17, 32
   weak 23, 24
propagation vector, *see* **k**
pulsed magnets 95, 107
purification of metals 131

quantisation of Hall resistance 147–50
quantum Hall effect 143–52
quantum oscillations, *see* magnetic quantum oscillations
quantum point contact 161–4
quantum well 68–70
   applications 70
   optical properties 69–70, 101–2
quasi-one-dimensional 78, 82, 190
quasi-two-dimensional 78, 103, 107–8, 143, 198
quasiparticles 112–4

reciprocal lattice
   definition 17
   unit cell of 17, 167
reciprocal lattice vectors 17
   algebraic expressions for 21
reciprocal mass tensor 47
recombination 155
reduced mass 193
relaxation-time approximation 2, **3**
   electrical conductivity 4, 14, 120–2
   Hall effect 6, 133–5
   magnetoresistance 135–8
   mobility 128
   plasma frequency 15
remarks, pious vii
resistivity (*see also* electrical conductivity)
   measurement of 194–9
   tensor 137–8, 145–50, 198
Ruddlesden–Popper oxides 80
Ruthenium oxide 130
Rutherford scattering 125, 128

sample geometries, for resistivity measurements 196
saturation 60
scattering (of electrons)
   absence of 117–19
   Bloch wavefunctions and 45
   by electrons 125–7
   by impurities 125, 128
   by phonons 122–5, 129

scattering (of electrons) cont.d
  in very pure metals 45
  relaxation-time approximation 2, 13, 16
scattering rate 2, 13, 16, **117–29**
  difference between thermal and electrical conductivity 120–2
  measurement using magnetic quantum oscillations 97
  temperature dependence, in metals 121–5
Schottky barrier 104
Schrödinger equation
  for a periodic potential 18–19
semiconductor
  alloys 65-6
  bandstructure calculations 63–4
  carrier density 58–62
  definition 49
  degenerate/nondegenerate 62
  devices 63–4
  direct gap 53–4
  electrical conductivity 127–9
  general descriptions 49–64
  heterojunctions 73–4
  indirect gap 51
  law of mass-action 56–9
  quantum wells 68–72
  thermal population of bands 56–9
  versus insulator 62–3
semimetal 73, 135
Shockley–Haynes experiment 156–8
Shubnikov–de Haas effect/oscillations 93, 96
Si (silicon)
  band gap 36, 37, 50–1
  bandstructure 50–1
  Brillouin zone and crystal structure 37, 49
  constant energy surfaces 52
  effective masses 100
  hydrogenic donors 60
  intrinsic carrier density 59
  MOSFETs 143
  tight-binding model of 36–7
  velocity saturation 158–9
Si–Ge superlattices 72
Skin depth 98
Sn (tin) 36–7
Sommerfeld model 7–13
  assumptions 7
  electrical conductivity 14, **119–27**

Fermi energy, velocity and wavevector 8, 9
heat capacity 11–12
magnetoresistance 135–7
scattering, mystery of 45
successes and failures of **13**, 135–8
thermal conductivity 119–27
space lattices 170–1
spacer layer 73–4
specific heat, see heat capacity
spin 173–4
spin-density wave (SDW) 82
spin-orbit split-off band 53
spin splitting 148–50
Sr_2RuO_4 (Strontium Ruthenate) 95, **106**
statistical weight 175
substrate 66–8
subband
  heterojunction 73, 143, 151–2
  quantum well 68–70, 83–4
superconducting magnets 94, 98
superconducting switch 131
superconductivity 77, 80, 81, 107, 111, **201–4**
superlattice
  buffer 68
  definition 71
  type I 71
  type II 71–2
  type III 72–3
surface reconstruction 103, 106
susceptibility 95
symbols, tabulation of 205–8
system 176

temperature, definition of 176
thermal conductivity
  definition 4–5
  electronic, Drude result 5
  temperature dependence for metals 119–27
  typical values for metals 5
thermal voltages, thermocouple 195
theological literary reference 45
thermopower
  of a metal 14–15
Thomson coefficient 14–15
tight-binding model 32–40
  comparison with nearly-free electron model 41
  effective mass 35, 47, 48
  Fermi surfaces 39–40

Hamiltonian 33
  transfer integrals 34
  wavefunctions 32
transfer integral
  and effective mass 35
  definition 34
  in organic metals 76–8
  pressure dependence 109–11
  relationship to underlying atomic orbitals 35
transition metals
  bandstructure 37–8, 48
  resistivity 126–7
  thermal conductivity 126
translation vector 165–6
translational symmetry 8, 16
  electronic potential 16
  k-space 17, 19, 24
  lattice 16
transverse magnetoresistance 136
trap 129, 155
two-dimensional electron systems (see also quasi-two-dimensional) 143–5
  conductivity and resistivity tensors for 145–50
two-wire method 194–5
type I, II and III superlattices 71–3

umklapp scattering 124

valence band
  definition 49
Van Hove singularities 105
vanishing resistance 147–50
variable-range hopping 130–1
velocity, of a band electron 42, 78, 120, 139
vertical process 122

warping 143
weak-coupling limit 202–3
Wiedemann–Franz ratio (Lorenz number)
  definition 5
  Drude value for 5
  experimental behaviour 6
  temperature dependence of 120–5
Wigner–Seitz cell 167–8

zinc-blende structure 37, 49